GEOLOGISTS' ASSOCIATION
1858
THE CURRY FUND

THE [illegible]

CHANNEL ISLANDS

by Clive Bishop,
David Keen,
Stan Salmon,
and
John Renouf

Guide Series Editor;
J. T. Greensmith

CONTENTS Page

PREFACE

The Channel Islands belong geologically as well as geographically to nearby Lower Normandy and Brittany - known collectively as Armorica. The pre-Palaeozoic, and much of the Palaeozoic geology of this Armorican province, differs fundamentally from that of the Cornubian massif of Britain. The two provinces have very different histories, the result of palaeogeographical separation. The distribution of the bedrock formations of Jersey can be seen in Figure 1.

Andrews (1976) established a framework of Rb-Sr and K-Ar isotopic ages which, although modified by later workers, indicated that the rocks of northern Armorica can be subdivided into two main groups: (1) a crystalline basement of restricted outcrop yielding isotopic dates older than 900 Ma which is referred to as the Icartian (± 1800 – 2000 Ma for the type Icartian on Guernsey) with other outcrops in northern Armorica and less certainly on Sark and Alderney and (2) a group of rocks comprising a series of metamorphic and igneous rocks (Guernsey, Sark and Alderney), low grade metamorphic sediments (Jersey and Guernsey), plutonic intrusions (all main islands), volcanic rocks (Jersey), a conglomerate (Jersey) and sandstone (Alderney) all of which belong to a major episode of plate tectonic subduction and mountain building known as the Cadomian Orogeny (Bishop *et al.*, 1975). Rocks of this second group are widespread in Armorica.

The events of the Cadomian cycle spanned the period from c. 700 Ma to nearly 400 Ma ago (see D'Lemos, *et al.*, 1990 for extended coverage of the Cadomian orogeny and Miller, *et al.*, 2001, for U-Pb dates for Jersey). It is agreed by most geologists working in northern Armorica that the Cadomian cycle can be related to an Andean Type subduction zone which dipped to the southeast under the 'northern' margins of the Gondwana continent. The extent and amount of underlying Icartian basement in the immediate area during the development of the associated volcanic basinal arcs is in dispute although U-Pb zircon ages of 2645 Ma from detrital grains in the Brioverian of L'Étacq suggest derivation from Icartian sources (Miller, *et al.*, 2001). However, in general it is accepted that the extensive Brioverian sedimentary and related volcanic rocks of northern Armorica were deposited in submarine fan environments. The widespread calc-alkaline plutonism which characterised the middle to late Cambrian evolution of the Cadomian zone in the Channel Islands and nearby Normandy was preceded by localised but important syntectonic plutonism, e.g. Guernsey's Perelle Quartz Diorite (U-Pb zircon age of close to 611 Ma, Samson & D'Lemos, 1999) and amphibolite facies metamorphism, e.g. the northern Channel Islands as reported by Dallmeyer *et al.* (1991, 1992, 1994) and D'Lemos *et al.* (1990). The less deformed and weakly metamorphosed Brioverian sedimentary and volcanic rocks of Jersey contrast markedly with the more deformed rocks of the other islands and would seem to indicate that the two areas were more widely separated in Cadomian times than now. U-Pb dates published by Miller, *et al.* (2001) indicate that at least part of the Jersey Shale Formation was deposited after *c.* 587 Ma. Dates obtained from the Anne Port Rhyolite

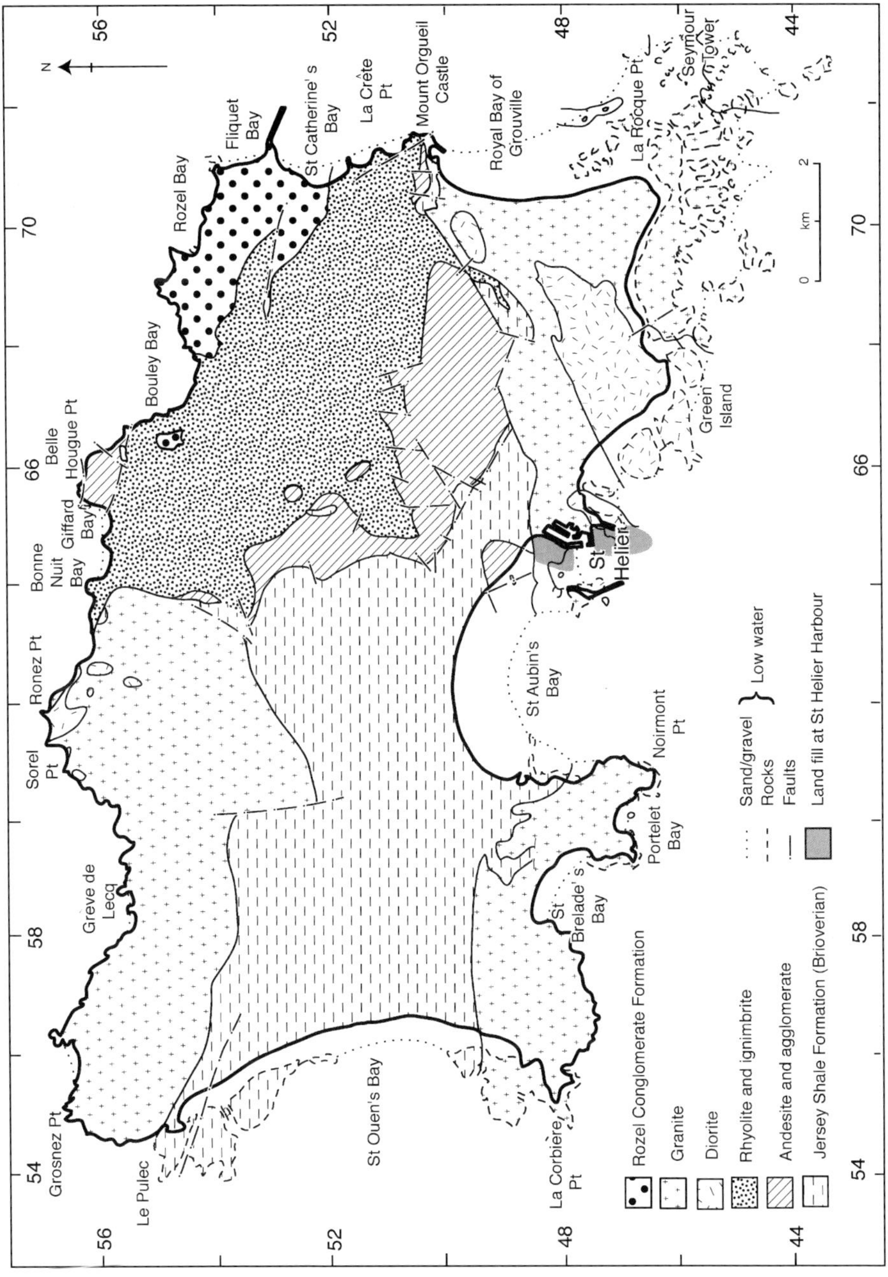

Fig. 1 Geological map of Jersey. (IPR/15-29C British Geological Survey.©NERC. All rights reserved)

indicate formation as sub-aqueous volcanics around 583 Ma. Both the Brioverian and the volcanics were folded prior to the emplacement of the Southwest Granite at *c*. 580 Ma.

After the cratonisation of northern Armorica, important spreads of fluvial to deltaic sandstones (Alderney and Cap Fréhel/Erquy), a series of fan conglomerates (Jersey's Rozel Conglomerate) and the extensive shallow water marine ortho-quartzites of the *Grès Armoricain* were deposited. The latter is present along the margins of the Gondwana continent from the Channel to Portugal and North Africa, though not preserved anywhere in the islands.

The final manifestations of the Cadomian Orogeny are thought to be a plutonic event which formed the Jersey Northwest Granite at *c*. 480 Ma and which is petrogenetically unrelated to the early Southwest Granite plutonism, and a series of dykes emplaced before *c*. 425 Ma ago post-dating the *Grès Armoricain* and the regional stabilisation of the Gondwana continental margins in the area. Following this event the islands have no rock record until the events and deposits associated with the Quaternary. Elsewhere in Armorica and its immediate surroundings there is a more tangible record of later Palaeozoic, Mesozoic and Tertiary times and, not to be overlooked, there are the Mesozoic and early Tertiary rocks on the sea bed between, and north of, the islands. Additionally, and affecting our understanding of the islands, is the rapidly increasing knowledge of middle to late Tertiary deposits of Armorica and their associated shorelines.

Pleistocene sediments in the islands (for Jersey see Figure 2) and in adjacent Armorica consist of marine deposits (raised beaches), terrestrial deposits (periglacial head and loess, peat, silt and blown sand) with erosional evidence in the form of marine notches, platforms and other physical features. Peats, alluvial silts and blown sand characterise the deposits associated with the Flandrian (Holocene) rise in sea-level. Considerable uncertainty still attaches to the age of many of these deposits and features within the islands. They are often found isolated one from the other and lack stratigraphical control either from their lack of context, i.e. they are neither preceded nor succeeded by datable deposits or features, or by the absence of internal evidence, e.g. fossil content or susceptibility to other forms of dating. The considerable advances made in unravelling the complexities of the continental Armorican successions (Somme, *et al*., 1995) and those of southern England – though not without controversy – have not transferred well to the islands. Their chief contribution is to warn that the successions in the islands may be equally complicated.

The oldest deposits dated at the present time are the basal temperate silts and sands at the base of the La Cotte de St Brelade cave succession which are older than 200 ka (Callow & Cornford, 1986), but raised beaches of the St Helier Formation associated with a sea-level *c*. 30m above the present are likely to be even older (Keen, 1978a, 1995). This is to be viewed within the Armorican context of such a cave site as Menez Dregan I in southern Finistère with sediments tentatively dated to *c*. 1 Ma (Monnier *et al*., 1997). Other dated elements are the important 8m raised beach of the Belle Hougue Member at the type site of Belle Hougue Cave where amino-acid and uranium series dates of 121 ka have been determined (Last

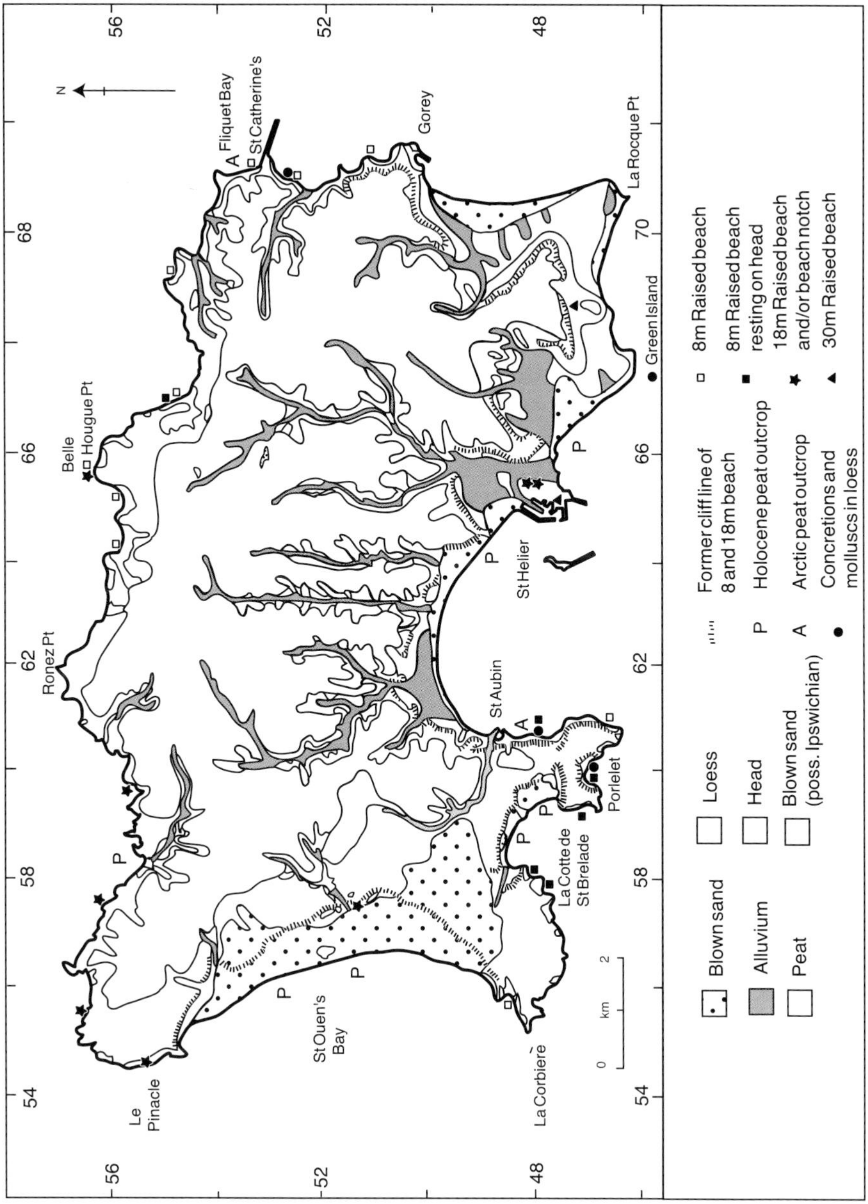

Fig. 2 The superficial deposits of Jersey. (IPR/15-29C British Geological Survey. ©NERC. All rights reserved).

Interglacial, Oxygen Isotope [OI] Sub-stage 5e). Compacted peat deposits of the Fliquet Arctic Bed and at St Aubin almost certainly date to the onset of cold climate conditions early in the last glacial stage, the Devensian. Other deposits in Jersey, notably those at La Cotte de St Brelade, Belcroute, Portelet and Green Island, have well attributed sequences together spanning from the late Middle Pleistocene to near the end of the Devensian and containing evidence of a range of climatic fluctuations.

The age of other marine sediments, with height ranges between 18m and 8m above mean sea-level, remain uncertain in most cases though a considerable number of the 8m beaches of the Belle Hougue Member are certainly attributable to the last interglacial, the Ipswichian, and are commonly overlain by variable thicknesses of periglacial head and loess. In contrast, the planation platform associated with the present tidal exposures in Jersey, around and among the other islands and along the adjacent French coasts in general, is a composite feature which may have been initiated possibly tens of millions of years ago (Renouf, 1993).

The extent and succession of the Holocene deposits in Jersey is known in some detail and stratigraphy based on a ^{14}C chronology was published by Jones *et al.* (1990). It is clear from the dates in this work that the maximum of the Flandrian Transgression was reached around 4000 BP. Peat and alluvium began to be deposited *c.* 8000 BP, while most of the blown sand of western and southern coasts is younger than 4000 BP.

The rocks of the island are well displayed in coastal exposures, and the clean, wave-polished surfaces are ideal for studying rock relationships and textures. These surfaces can be, and in places have been, spoiled by indiscriminate hammering. **All who use this guide are urged to use their hammers sparingly and in such a way as to preserve the geological amenity.** In certain places hammers should not be used at all, and these are specifically indicated in this guide. Further note should be taken of the designation of the whole of the southeastern foreshore of Jersey as a RAMSAR (wetland) site with quite specific obligations on all users. Several sites on the included itineraries are located within the boundaries of this designation. Further information, or a reminder, is included within the descriptions of the relevant geological sites. For further information contact the Jersey Planning and Environment Committee.

Take particular care to find in advance the times of low and high water. The tidal range in the Channel Islands, and in Jersey in particular, is much greater than elsewhere in Britain and the rate of rise of the tide is very rapid and is accompanied by very strong currents. Never venture on to rocks exposed at low tide unless thoroughly familiar with the disposition of the reefs and the tidal runs. A booklet of tide-tables is published annually by the *Jersey Evening Post* and this newspaper daily includes tables giving tidal predictions for a few days ahead. Such tables can also be consulted at the Harbour office.

Map references given in this guide are to the Universal Transverse Mercator grid on the 1:25000 Ordnance Survey/States of Jersey map of the island. The British Geological Survey Channel Islands Sheet 2 1:25000 (Solid and Drift) map (1982) covers the whole of the geology of the island. Copies of these maps may be obtained from Ordnance Survey agents or from the Museum of La Société Jersaise in St Helier.

LIST OF FIGURES

THE SOUTHEAST IGNEOUS COMPLEX

This complex extends from Elizabeth Castle and St Helier to Gorey. It comprises dioritic and gabbroic rocks cut by granites of several - though all Cadomian - ages. Significant elements are the older Dicq granite, the younger Le Hocq-La Rocque-Gorey granite and the Fort Regent granophyre. The complex is intruded by a swarm of basic, acid, composite, and lamprophyre dykes of several ages. The best exposures are on the inter-tidal reefs, which at low tide extend 1.5km or more beyond the coast. **It cannot be too strongly emphasised that to venture on to these reefs without knowledge of the tides and tidal runs is extremely dangerous. Many visitors and even local residents have lost their lives in calm weather on this coast by being cut off by the incoming tide.**

ITINERARY 1
ST HELIER

Elizabeth Castle (half tide to low tide) should be visited on a falling tide by walking across the causeway leading from the West Park slip opposite the Grand Hotel taking care to return in time to avoid being cut off by the incoming tide. Although amphibious vehicles are usually available for the journey, more geology can be seen by going on foot.

Time allowing, a complete circuit westwards round the castle to the Hermitage and back along the eastern foreshore is recommended. If time is short or if the priority is the diorite/granophyre relationship then go southwards to the Hermitage on the quicker eastern side and work your way back the same way.

It is possible that the layered diorite was intruded by a co-existing granophyre magma (Shortland *et al.* (1996)). Alternatively, the diorite might have been emplaced into the granite magma (Weibe, 1993). Following commingling of the two magmas, physical and chemical interactions produced a variety of textural and mineralogical changes. Later granophyre incursions disrupted the diorites. The result is two major and distinct patterns: (1) where the layered diorite has maintained its original form and (2) where it has been largely broken up into fragments. The complex thus allows ample opportunity to compare the fragmentation of diorite in both the magmatic and solid state. Bishop & Key (1983, 1984) proposed an alternative, largely metasomatic origin for the diorites from gabbros, with contact relationships interpreted as being due to recrystallisation and rheomorphism of the gabbro.

A first introduction to the diorite/granophyre relationships is seen west and south of the castle entrance below the first curtain wall (Figure 3, location 1,

639479) where bluish black, rather fine-grained diorite is intruded by veins of pink granophyre, though a lens is needed to see the micrographic texture. The granophyre contains angular fragments, or enclaves, of diorite, the edges of which may be sharp and angular, lobate, or gradational. The granitic rock in contact with the diorites is commonly grey and has a greater proportion of coloured minerals than the pink granophyre. This grey rock is a marginal hybrid facies produced by interaction between the diorite and granophyre. This area represents remnant and broken sheets of the diorite.

About 200m southwest of the castle entrance (638478), the diorite gives place

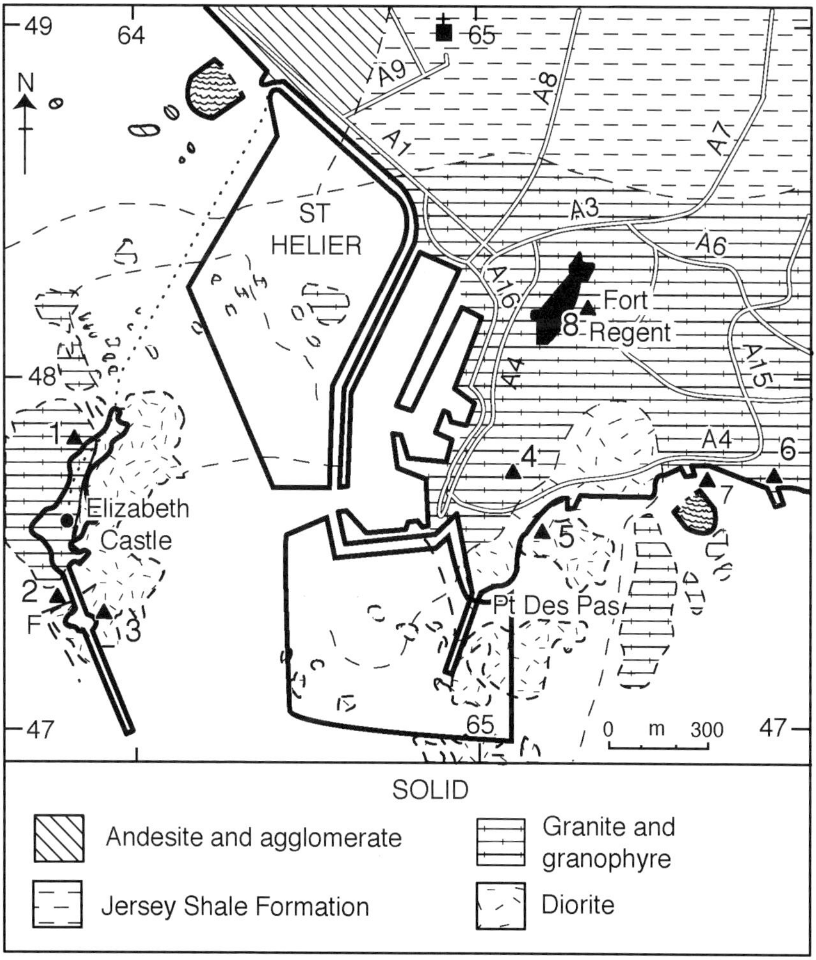

Fig. 3 Geology of the St Helier/Elizabeth Castle area. (IPR/15-29C British Geological Survey. © NERC. All rights reserved).

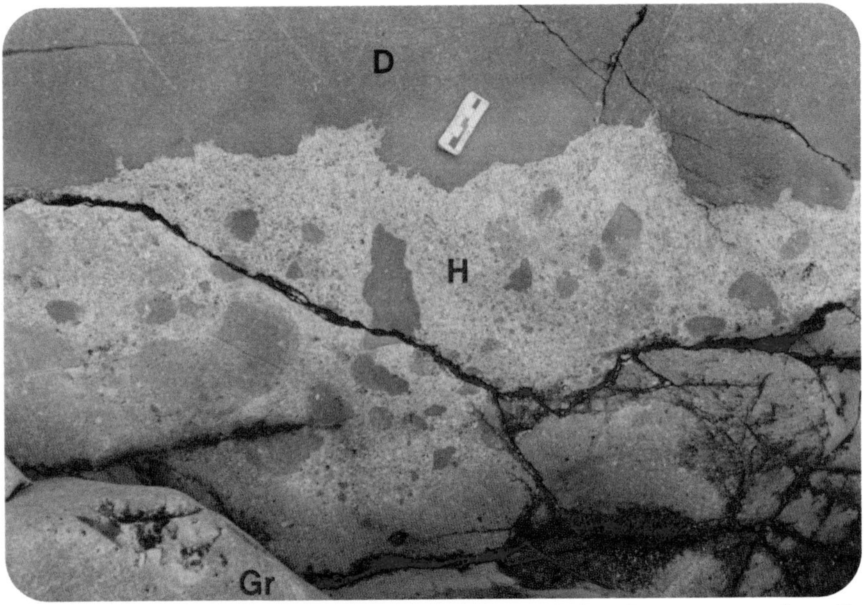

Fig. 4 Magmatic interactions at Elizabeth Castle, below the eastern castle wall. Diorite (D) is separated from granophyre (Gr) by a pale hybrid zone (H) within which are enclaves of the diorite. The diorite has a crenulate, fine grained (chilled) margin, indicating that it and the granophyre were present as coexisting magmas. The hybrid was formed by chemical interaction between them.

to close-jointed granophyre which continues southwards until the diorites appear again near the Hermitage, south of the castle. Several dykes, trending about 20° north of east, are exposed on the west side of the breakwater linking the castle with the Hermitage. A 5m composite dyke (Figure 3, location 2 at 638474) is worthy of particular note. The doleritic margins and porphyritic microgranite interior are of approximately equal thickness, in contrast to most other composite dykes in Jersey in which the microgranites are much thicker than the basic margins. Small veins of microgranite extend into the flanking dolerite and there are dolerite inclusions in the microgranite.

The Hermitage Chapel stands on a rock pinnacle immediately east of the breakwater (639473, location 3 on Figure 3) and it is here and in the immediate surroundings that the intrusion of granophyric magma into, and sub-parallel with, the low dipping layered dioritic magma can best be examined.

Two conspicuous outcrops immediately west of the Hermitage superbly display parts of the sheeted complex. Sheet margins are almost universally crenulate often at millimetric scales and have conspicuous fine-grained margins, interpreted as chilled margins, about 1cm wide. In places, straight-sided veins of granophyre cut the sheet margins and it is considered that this represents a penetration of very viscous or even solid diorite by still mobile granophyre. The rock face immediately below the Hermitage consists of a further hybrid rock produced by extensive fragmentation of diorite, down to

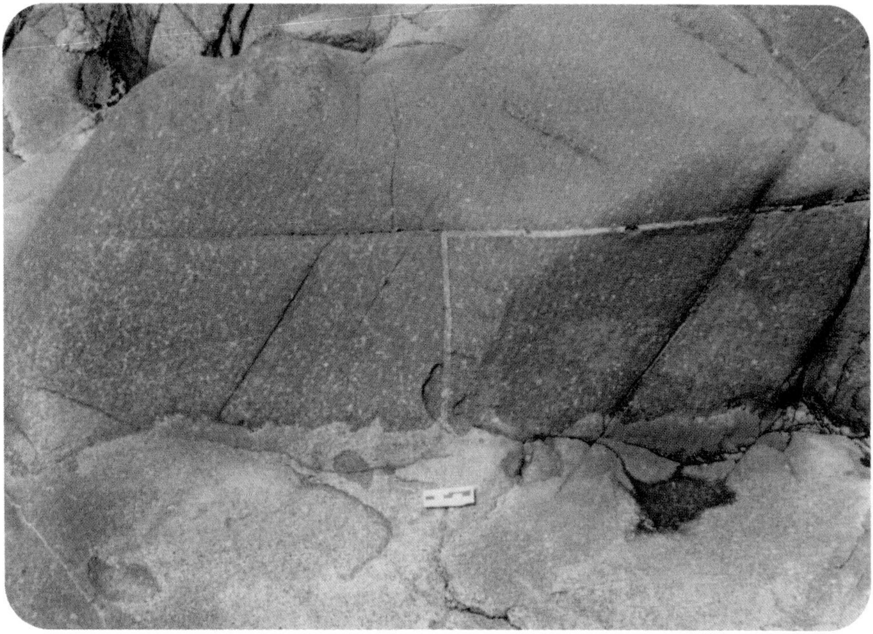

Fig. 5 Relationships at Elizabeth Castle, below the eastern castle wall. The irregular contact between diorite (top) and granophyre indicates that they were initially present as coexisting magmas. The diorite crystallized more quickly than the granophyre and then fractured, the fracture being filled with granophyre magma.

millimetre-scale clots. Several diorite varieties are recognisable on this face.

Several basic dykes cut the diorites. The most interesting is a 5m wide porphyritic variety with greenish plagioclase phenocrysts which is cut by a 30cm wide aphyric dyke at the northern end of the small quarry opposite the Hermitage.

On the eastern reefs, from below the Hermitage back to the Castle entrance, the sheeted diorite has been largely broken up into fragments in a manner suggestive of stoping but the precise mechanism and the physical nature of the diorite and granophyre at the time remains uncertain (Figures 4 & 5). The fragments of diorite show all stages of break-up from clean fracturing (angular pieces) to almost complete absorption (mafic clots in the granophyre). Low outcrops immediately below the castle wall consist of further intact fragments of the sheeted complex.

Around the bathing pool at the end of the causeway at West Park, andesitic lavas and agglomerates are exposed on the isolated reefs of rock.

St Helier and South Hill. Fort Regent is built on a hill of granophyre similar to that of Elizabeth Castle. From the town, walk along Pier Road to South Hill (650477) observing the scarped cliffs of granophyre beneath the wall of Fort Regent. Near the western end of the southern face of South Hill, the granophyre is intruded by basic dykes, one of which thins upwards. At the eastern end of this section there are two hornblende-mica lamprophyre intrusions which are cut and displaced by gently dipping shears (Bishop, 1964a). **The rocks here are unstable and should not be hammered**. Further east, at the entrance to the gardens that now occupy a disused quarry, there is a 3m weathered mica lamprophyre (minette) dyke (Smith, 1936). The minette cuts a basic dyke but the contact is hidden behind bushes.

Cross to the opposite side of the gardens and from the steps look back at the eastern face of South Hill. At the cliff top and to the north (right) of the outcrop of the minette there is an exposure of a raised beach at 35m, which is hidden by a clump of ivy at the time of writing. (Figure 3, location 4). This outcrop is not easy to make out. Rounded cobbles and pebbles of local derivation occupy a small hollow in the granophyre and are overlain by angular, yellowish loessic head. These deposits are of uncertain age although their height above sea-level may suggest a date in an interglacial of Middle Pleistocene age (Keen, 1978a; 1993; 1995). Deposits at a similar height in the Boxgrove-Slindon area of Sussex (Bates *et al.*, 1997) have been dated to late in the Cromer Complex (Middle Pleistocene) by mammalian biostratigraphy. A similar age (OI Stage 13 or older) would be probable for South Hill. However, beach levels at such altitudes possibly reflect neotectonic uplift and dating by level alone is suspect.

THE FORESHORE FROM SOUTH HILL TO GROUVILLE BAY ARE PART OF A DESIGNATED RAMSAR (WETLAND) SITE TO WHICH SPECIFIC RESTRICTIONS APPLY.

From South Hill follow the path between the old barracks and the power station to the shore (on a falling tide) at Pte des Pas, now reclaimed land (651474). There are different forms of fill but the most conspicuous is the outer armourstone apron (rip-rap) comprising large blocks of rock derived mainly from sources outside the island. Continue along the path to the point at which it turns sharply left near to a shelter. The La Collette sill (Bishop, 1964b) is exposed in diorite on the rocks uncovered here at mid-tide (Figure 3, location 5). The sill dips gently northwards and can be recognised by a prominent band of pink microgranite 30cm thick near its top. The upper chilled contact of the sill is exposed about 1m above this microgranite and the altered dolerite of the sill is remarkably similar in appearance to the surrounding diorites. The sill is about 10m thick and is cut by several ENE-trending basic dykes.

The best section through the sill is exposed on the most easterly reefs, though the base of the sill, in places considerably epidotised, can be seen by the small rocky pier which extends seawards from La Collette walk. The sill was formerly exposed immediately south of Victoria Pier where it was seen to intrude the hornblende-mica lamprophyre dykes of South Hill.

Dicq Rock and Havre des Pas (high tide and falling). Follow the coast road to the Dicq (659478) where a slipway leads to the beach (Figure 3, location 6). Rocks by the slipway are of porphyritic adamellite, the Dicq granite. In places a weak, nearly vertical and north-south alignment of potassic feldspar megacrysts is discernible. The granite also contains rounded, dark inclusions, many of which contain potassic feldspar megacrysts similar to those of the parent granite. On the western side of the promontory the granite is cut by a basic dyke about 3m wide which contains several blocks of granite (Casimir & Henson, 1949). The rock bears a memorial tablet to Victor Hugo who liked to sit here during his period of exile in the island.

Cross the sandy beach west of Dicq rock to the low rocks northeast of the bathing pool (Figure 3, location 7). Dicq adamellite is again exposed but the most northerly exposures are of Longueville granite — a variant of the Dicq mass which lacks the abundant mantled potassic feldspar megacrysts and is characterised by yellowish plagioclase feldspars.

Snow Hill. The cutting at Snow Hill in St Helier was originally part of the scarping carried out for the Fort Regent defences; subsequently it has been used as the terminus of the Jersey Eastern Railway, a bus station with turntable, and a public car park. It now allows access to the Fort Regent Cavern which is a large underground storage chamber blasted out (1995-1998) of the Fort Regent granophyre beneath the Fort to serve as an emergency storm water holding tank for St Helier. On the eastern side of the cutting to the south of the walling there is an exposure of a 1m columnar-jointed dolerite dyke (654482)(Figure 3, location 8). The dyke is not exposed on the western side having been displaced by a fault inferred to run along the cutting. About 100m further north, and on the western wall, a similar columnar-jointed dyke is exposed and has no counterpart on the eastern side: this could be the displaced part of the dyke. The access portal to the cavern is some 20m north of this dyke.

Immediately south of the dyke in the eastern wall there is an exposure of the 18m raised beach (Railway Terminus Member) now rather overgrown by vegetation. Rounded cobbles and pebbles in a coarse sandy matrix rest on smooth, water-worn surfaces of granophyre. The geomorphological relationship of this raised beach to that at 30m on South Hill indicates that it is not the same deposit. Beach remnants in Guernsey (Keen, 1978a) suggest that beaches at around 20m above mean sea-level are intermediate in age between those at 30m and those at 8m, but the exact age of such unfossiliferous remnants is very difficult to determine. Consideration of beaches in this intermediate position in the Channel area as a whole (Keen, 1993; 1995) suggest a Middle Pleistocene age, and the complex sequences of beaches on the Sussex coastal plain described by Bates *et al.* (1997) would allow an age in either OI Stages 11, 9 or 7 for deposits at this height. Amino-acid geochronology on shells from raised beaches at Portland (Davies & Keen, 1985) and Torbay (Mottershead *et al.*, 1987) suggest that levels as high as 18m should be older than OI Stage 7, but exposure and high tidal range in Jersey make correlation with OI Stages 7, 9 or 11 inexact. **Please do not hammer or collect from this exposure.**

ITINERARY 2
GREEN ISLAND AND LE NEZ
(half tide to low tide)

THE FORESHORE FROM SOUTH HILL TO GROUVILLE BAY IS PART OF A DESIGNATED RAMSAR (WETLAND) SITE WITH SPECIAL RESTRICTIONS REGARDING REMOVAL OF SPECIMENS.

Descend to the beach at Green Island slip and turn west to Le Croc Point (673462). At Le Croc, diorite is intruded by numerous dykes displaced by one another and by faults (Figure 6, location 1). The earliest member of the suite is a 10m hornblende lamprophyre exposed at the point of the sea-wall. This dyke has basic margins, on the inner side of which is porphyritic microgranite which is in contact with the central hornblende lamprophyre. The contacts between the last two are suggestive of magmatic interactions between co-existing magmas. This intrusion is displaced by a second, 4m composite dyke comprising porphyritic microgranite with basic margins which is itself displaced by faults. The third phase of acid dyke intrusion was a porphyritic microgranite without basic margins. The intrusion forms two parts *en échélon*, the eastern portion being connected to the western part by a cross member that displaces all earlier dykes. Several later phases of basic dyke intrusion, themselves displaced by faults, cut across all earlier dykes.

At low tide cross the sandy gully to Green Island, a remnant of the former loess cover of the shore platform (Figure 6, location 2). The rocks at and around Green Island show many features characteristic of the diorites of the Southeast Complex. South of the island there are fine examples of inclusions of diorite in granodiorite and quartz-diorite. The inclusions range from sharply angular fragments to shadowy xenoliths that were thoroughly plastic and 'flowed' along with the surrounding rocks. At the southeastern part of the island there are some dense, black, fine-grained xenoliths with crenulate margins. These differ from the more common dioritic rocks and they are probably fragments of early synplutonic dykes caught up in the intrusion. Northwest of the island there are exposures of layered diorite similar to, but not as perfectly developed as, those at Le Nez.

The igneous rocks at Green Island are overlain by 0.25m of raised beach gravel and then by up to 3m of loess. The top of the islet is capped by up to 2m of blown sand. The loess is largely decalcified and on the south side of the island exposures, now covered over by conservation measures, exhibited *limon-à-doublets* banding formed during decalcification. At the base of the section at the eastern end of the north face of the islet, the loess is still calcareous and contains concretions (*Lößmännchen*) up to 10cm long. In and around these concretions a sparse non-marine molluscan fauna of *Pupilla muscorum* (Linné) and *Oxyloma pfeifferi cf*

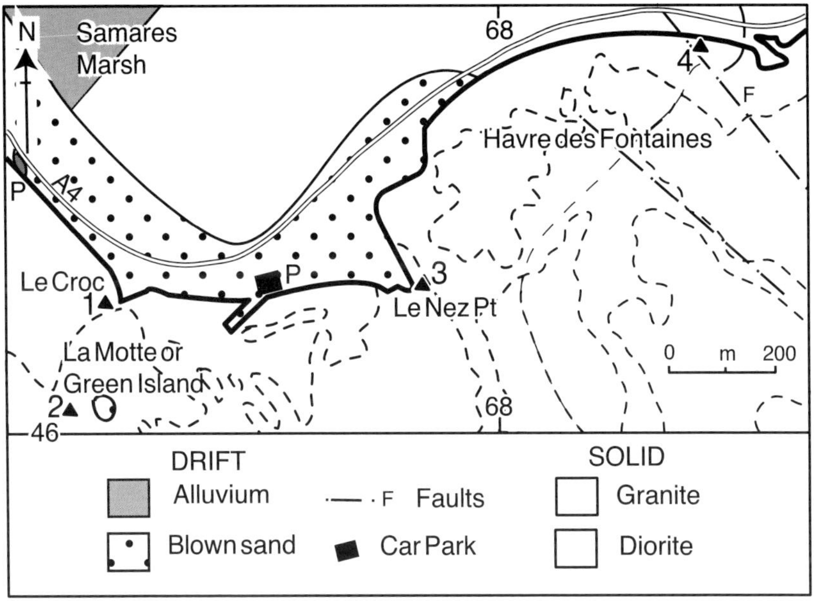

Fig. 6 Geology around Green Island and Le Nez. (IPR/15-29C British Geological Survey. © NERC. All rights reserved).

schumacheri Andreæ (Keen, 1978a), which indicate cool, open habitat conditions during the deposition of the loess.

The blown sand is of disputed age (Neolithic to Medieval (Patton, 2002)) and has mostly been destroyed by marine erosion. In the blown sand, middens with marine mollusca (chiefly *Patella vulgata* Linné and *Littorina* spp), animal bones and artefacts - the food debris of the cemetery builders - are found and span a considerable age. Digging or collecting should not be attempted in these layers.

THE EXPOSURES ON THE ISLAND ARE NOT TO BE DISTURBED IN ANY WAY EXCEPT WITH PRIOR AND SPECIFIC PERMISSION OF THE OWNERS, THE SOCIÉTÉ JERSIAISE. IN PARTICULAR NO CONCRETIONS (*LÖßMÄNNCHEN*), FOSSILS, POTTERY OR STONE IMPLEMENTS SHOULD BE TAKEN.

Walk eastwards to the small promontory of Le Nez (678462), to examine layered diorites exposed on clean, wave-polished surfaces (Figure 6, location 3). These exposures should not be hammered: they are worthy of preservation.

The layered sequence has a NW-SE strike and dips at about 70° to the Northeast. The layers are about 1m thick and each passes upwards gradually from dark meladiorite - which, for the first 10cm or so, is notably finer-grained than the overlying diorites - into progressively lighter diorites. The uppermost member of each sheet is quartz-diorite which often contains elongate amphiboles showing flow alignment. These appinitic quartz-diorites frequently project upward into the dark base of the next unit, and in places form small diapirs penetrating the darker, overlying diorites.

The lower, layered diorites give place upward to layered poikilitic hornblende gabbros containing relict pyroxene. The gabbros are recognised by their spotted and flecked appearance. The spots, up to 3cm across, are of poikilitic amphibole. The gabbros also show fine-scale layering comprising several 'centimetre-scale' layers of light and dark diorite closely associated with a pegmatitic facies showing, at its base, crescumulate textures. **Please do not hammer or collect from these exposures.**

The gabbros give place upwards to coarse appinitic rocks containing elongate amphiboles with feldspar cores (Wells & Bishop, 1955). The appinites are diorite pegmatites occurring as patches either along the layering, as elongate cross-cutting areas, or as large pockets which truncate the layering. Immediately west of Le Nez, there is a return to dioritic layers similar to those of the first exposures. The layering extends from the reefs at Grève d'Azette west of Green Island as far as the granite contact in Havre des Fontaines, a distance of nearly 2km. The strike and dip remain sensibly constant over this distance, so the vertical thickness of layered rocks is considerable even if allowance is made for the effects of faulting. Two important explanations have been offered on the genesis of these layered rocks. In the first (see Bishop & Key, 1983, 1984), recrystallisation and mobilisation of a layered gabbro consequent on the emplacement of granite created the varied dioritic complex now seen; in the second (e.g. Topley & Brown, 1984, Shortland *et al.*, 1996) the origin of the diorites is sought in their production by primary crystallization from a diorite magma. The complex has been much disrupted by faulting and dyke intrusion.

Eastwards from Le Nez across Havre des Fontaines the sombre grey diorites give place to pink granite. This is the younger granite of the Southeast Complex and it can be examined near Le Hocq (685466). The contact between granite and diorite is faulted and runs roughly NE-SW through Havre des Fontaines (Figure 6, location 4). The granite continues round the southeastern coast to Gorey, though there are further exposures of diorite on the reefs about a kilometre offshore near Seymour Tower (723457). **It cannot be too strongly emphasised that the walk to Seymour Tower should only be attempted by those who are thoroughly familiar with the reefs, the tide and tidal runs.**

The granite is rather uniform adamellite though, at La Rocque, it is coarser than at either Le Hocq or Gorey, and contains several pockets of pegmatite and veins of aplite. Inland of the coastal dune belt from La Rocque to Gorey lies a low-lying area centred on Grouville Marsh. Like other coastal lowlands in Jersey the marsh is a product of sedimentation, often of reworked loess, behind coastal dunes driven inland by the rising Flandrian sea-level between 10,000 and 5000 years ago. Sediments in Grouville Marsh are up to 8m thick and have yielded radiocarbon ages of 6130 BP at their base (Jones *et al.*, 1990).

EASTERN JERSEY
ITINERARY 3
MONT ORGUEIL to LA COUPE (half tide to low tide)

Mont Orgueil Castle (716503) stands on the most northerly outcrop of the younger granite of the Southeast Igneous Complex. Descend to the shore by the steps over the harbour wall at the landward end of the quay. The granite here is an alkali-granite composed dominantly of potassic feldspar and quartz with chlorite and biotite. Near to the quay at Gorey is a 1m dyke of purplish brown rhyolite with flow-banding parallel to its margins (Figure 7, location 1). This exposure was at one time interpreted as a remnant of extrusive rhyolite which had flowed over the eroded surface of granite and filled fissures and gullies within it, but this view is not substantiated by the field evidence. There are other intrusions of flow-banded rhyolite; one is near St Aubin (see page 21), and another intrudes pyroclastic rocks south of Anne Port.

Three intrusions of porphyritic microgranite are exposed at the easternmost point of the castle and a further one by the harbour wall. Porphyritic microgranites comprise a significant part of the dyke swarm of southeast Jersey and occur usually as ENE trending dykes up to 10m across, commonly with marginal 'dolerites' about 30-50cm wide. The dykes usually weather pinkish yellow and have a distinctive joint pattern which gives them a spiky appearance. The dyke closest to the castle wall, immediately beneath the low cliff, lacks basic margins but two others to seaward are composite, and all three dip steeply southeast. The margin of the central intrusion branches so as to include screens of granite.

On the rocks south of the sandy beach at Petit Portelet (716505) (Figure 7, location 2), there are exposures of a coarse mica lamprophyre (minette) dyke which weathers to form a hollow. This dyke is exposed also south of the castle. The rocks north of Petit Portelet are rhyolitic ignimbrites but the contact with the granite is not exposed: the junction is most probably faulted. There are good exposures of head up to 10m thick backing the beach at Petit Portelet, and also behind the buildings at the northern end of the harbour frontage at Gorey.

Anne Port to St Catherine's. From Petit Portelet northwards to Anne Port rhyolitic ignimbrites containing blocks of andesite and 'shale' are exposed and are cut by microgranite dykes and, at Jeffrey's Leap, by a rhyolite dyke with flow-banded margins (Figure 5, location 3). A 10m multiple basic dyke is associated with this rhyolite dyke. It is a rough scramble round these rocks and the traverse cannot be made above half-tide. Instead, take the path to the road at Petit Portelet and northwards along the B29 coast road from which there are views to St Catherine's Breakwater, the islands of the Écréhous reef beyond and, in clear weather, France.

Descend to the beach by the slipway at Anne Port (713509) and walk northwards. Note the use of gabions to stabilise the steep slopes at the north and south of the bay. Between Anne Port and La Crête Point there is a well exposed, northward-younging and dipping sequence of volcanic rocks comprising pyroclastic lava and debris flows, some

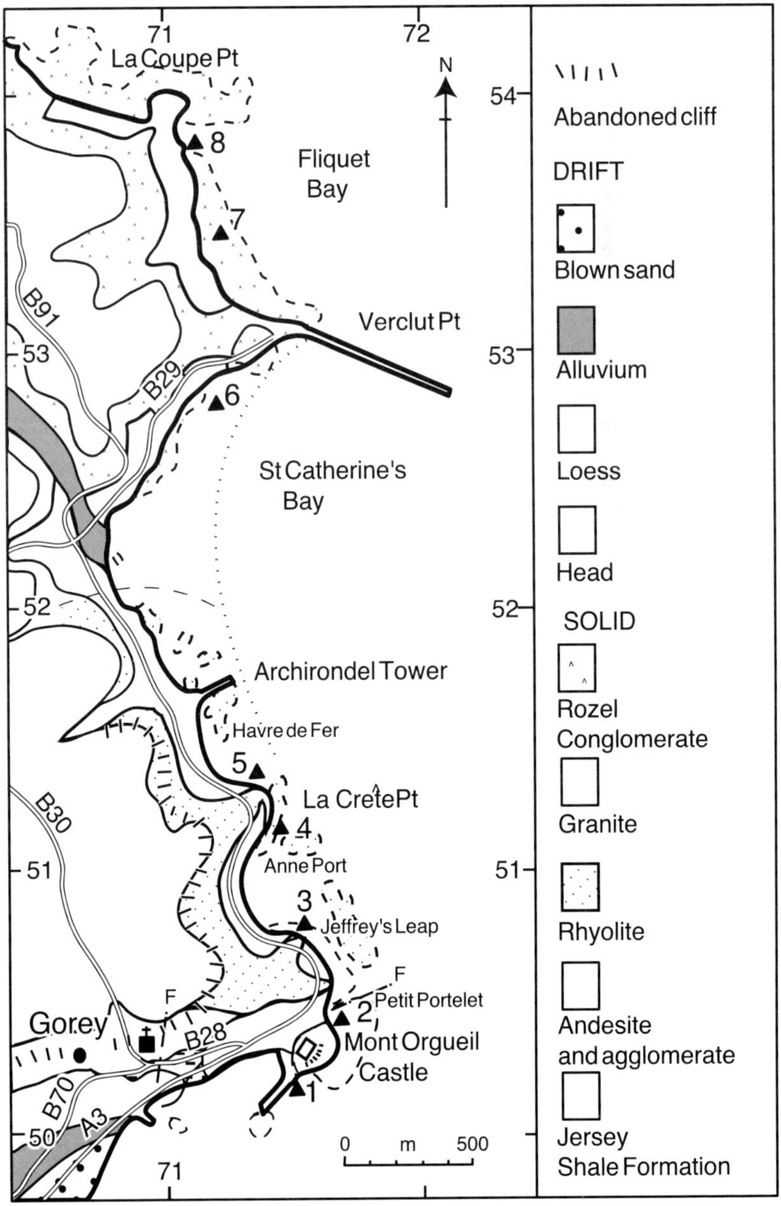

Fig. 7 Geology of the east coast of Jersey. (IPR/15-29C British Geological Survey. © NERC. All rights reserved).

Fig. 8 Autobrecciated top of a flow-banded rhyolite lava flow, north of Anne Port beach. The width of the field of view is about 40 cm.

quite spectacular (Figure 7, location 4). There are ten mappable units commencing at the shingle edge with a welded tuff of some complexity resembling flow-banded lava in parts and containing conglomeratic lenses up to more than a metre thick. A straight boundary to the north separates this tuff from a debris flow with clasts up to half a metre in diameter. The first of the rhyolitic flows follows but this wedges out away from the cliff and is followed by another debris flow and a vesicular andesite sheet, possibly intrusive. Above the andesite is another impressive autobrecciated rhyolite flow with blocks up to 60cm in size often showing folded flow-banding (Figure 8). Another very coarse debris flow follows and contains boulders up to 3m across. A thin flow-banded rhyolite and a thin debris flow separate the preceding coarse debris flow from a very thick, flow-banded and columnar jointed rhyolite that continues to the point and the angle of the headland. Do not spoil the weathered autobrecciated surfaces by hammering.

Exposures of rhyolite continue in Havre de Fer (713515) and columnar jointing in rocks exposed in the bay at low tide (Figure 7, location 5) can be seen from La Crête Point. These are the same flows with columnar jointing as those at Anne Port, repeated by faulting. The traverse can be continued at low tide into Havre de Fer by scrambling over rough and seaweed-covered rocks, by retracing one's steps to Anne Port (the easy route), or by taking the rough path upwards through the thickets to the road. The last exposures of volcanic rocks occur about 500m north of the Archirondel breakwater and Tower at 712517. They are predominantly ignimbrites, some containing bands of spherulites together with some water-laid tuffs and debris flows.

Northwards the volcanic rocks are overlain unconformably by the Rozel Conglomerate Formation, but the contact is not exposed on the coast. In times past various origins and ages have been ascribed to the conglomerate, but it is now accepted as a fan-conglomerate of late Cambrian or earliest Ordovician age, the bulk of the deposit comprising coarse, proximal material with very subsidiary finer-grained more distal layers (Went *et al.*, 1988; Went & Andrews, 1990).

There are good exposures of conglomerate at St Catherine's both in the disused quarry (712531) from which the blocks were obtained for the construction of the breakwater and on the shore north of the breakwater. Most of the pebbles and cobbles are of Brioverian sediment and of granite with yellowish feldspars. Between the slipway and the breakwater (Figure 7, location 6 at 711529), below the picnic spot, there is a gully high up on the foreshore which shows some finer grained sedimentary layers. At the entrance to the breakwater there is a large boulder of granite about 1m across - The Pebble so-called - that was obtained from the conglomerate. Notice that the large capstones of the breakwater are of grey granite with prominent white feldspar megacrysts, typical of many Cornish granites. The large setts backing the capstones, however, are of Jersey granite.

The conglomerate is in places cut by dolerite and hornblende lamprophyre dykes; one of the former is exposed on the shore about 200m east of the road fork at the commencement of the one-way road system around the old quarry and just east of the sailing club premises.

From the breakwater walk north along the coastal path or foreshore to Fliquet (712535) or drive around by road. In a wide east-west trending gully in the shore platform chance exposures may show up to 0.6m of gravelly peat (Figure 7, location 7). The flattened condition of pebbles and wood fragments in the peat shows clearly that it was compressed by the overlying head which now backs the gully, indicating that it marks an early episode in a cold stage during which the head was deposited. Pollen and insect remains from the peat, the latter visible with a hand lens, show that the base of the deposit was formed in a climate which allowed woodland dominated by *Pinus* (pine) and *Betula* (birch) to flourish, but the bulk of the deposit accumulated in treeless tundra perhaps like that of arctic Finland at present (Coope *et al.*, 1980). The age of this deposit is uncertain. Coope *et al.* assumed an early Devensian age on the premise that the head is Devensian, but modern views on the complexity of the Pleistocene coastal sequences in Jersey (Keen *et al.*, 1996) do not allow a conclusive attribution of age, and the Fliquet Arctic Bed may have been deposited at the start of any one of a number of cold phases from the late Middle Pleistocene.

At half-tide it is possible to walk northwards across the conglomerate exposures of the shore platform to La Coupe (710540), but at higher tides the road from Fliquet should be followed inland until La Coupe is reached. On the south side of the sandy cove south of La Coupe point there is an inlet floored by sand at half-tide, which is backed by a cave (Figure 7, location 8). The inlet and cave are eroded along an east-west composite dyke comprising a central hornblende lamprophyre, some 2m thick, flanked by thin marginal dolerites. A second pair of thin basic dykes is separated from the main intrusion by screens of conglomerate about 30-40m thick.

SOUTHWEST JERSEY
ITINERARY 4

ST AUBIN TO ST OUEN'S BAY

From St Aubin walk alongside the harbour and descend to the beach by the car park at the Bulwarks (607486). The rocks on the shore and stretching to St Aubin's Fort are hard, brown to grey alternating Brioverian siltstones and mudstones with coarser arenaceous layers up to 10cm thick striking roughly NNE-SSW and dipping steeply southeast. St Aubin's Bay has been formed by erosion of these sediments between the more resistant Southeast Igneous Complex and Southwest Granites.

Ar locality 1 (Figure 9), at mid-beach level, hollows in the Brioverian bedrock, intermittently exposed by movement of modern beach sediment, contain pockets of compressed organic mud up to 30cm thick. These muds have yielded pollen and insects indicative of sub-arctic conditions (Coope *et al.*, 1985). As at Fliquet, the compressed condition of the organic deposits indicates that they have been exhumed from beneath a cover of head. Although Coope *et al.* assumed an early Devensian age for the organic muds, the complexities of late Pleistocene stratigraphy in the area (Keen *et al.*, 1996) would equally allow an age at the end of OI Stage 7 or even earlier.

Walk south to the first promontory (Figure 9, location 2), on the southern side of which is a dyke of flow-banded rhyolite trending roughly east-west. A short distance further south at the back of the embayment there is a good section in Pleistocene deposits (Figure 9, location 3, Figures 10 & 11):

5.	1.4m +	Loess with *limons-à-doublets*
4.	15m +	Head composed of angular fragments of Brioverian sediments.
3.	2m	Well bedded medium sand with truncated soils and angular pebbles becoming common near the top.
2.	0.5m	Raised beach (8 m) largely composed of rolled granite cobbles but with some flints (Belle Hougue Member)
1.	1m +	Stony head of angular Brioverian fragments and loess with marine sand pockets in the base.

The contrast between the granite of the raised beach and the Brioverian fragments of the two heads is very well marked and indicates that a periglacial episode preceded the deposition of the 8m beach of the Belle Hougue Member. The age of this phase is uncertain but it is clearly pre-OI Stage 5. The occurrence of head below the raised beach indicates that the rock platform, which lies below the lower head, was already cut before the OI Substage 5e transgression, indicating an earlier sea-level around 8m, i.e. higher than that of the present beach (Keen *et al.*, 1996).

Continue southwards along Belcroute Bay across the contact between the Brioverian sediments and the marginal facies of the Southwest Granite. The contact

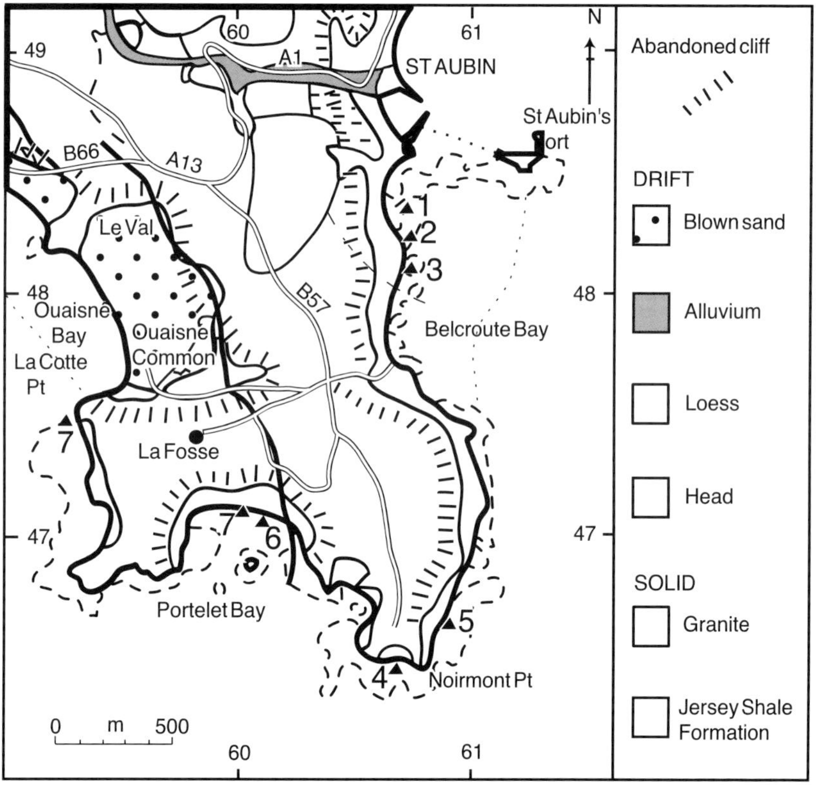

Fig. 9 Geology of the Noirmont peninsula. (IPR/15-29C British Geological Survey. © NERC. All rights reserved).

is not exposed but can be inferred to run in an E-W direction under the cobble beach between two low rock reefs. The sediments become tough and distinctly purplish towards the contact and contain small biotite flakes. The marginal granite is rather fine-grained and grey in contrast with the coarse, pink, biotite-hornblende granite that forms the bulk of the mass.

It is possible at low tide to scramble along the shore to Noirmont Point (607464) and examine the granite which is cut by several east-west trending basic dykes, up to 2m in width, of the Jersey dyke swarm, but care is needed to avoid being cut off by the tide. Alternatively, take the road to Noirmont Point and descend to the lighthouse (Figure 9, location 4). The granite is intruded by several basic dykes and, in a gully between the lighthouse (a converted Martello Tower) and the German ranging tower, there is a 1m north-south mica lamprophyre dyke which

Fig. 10 The Belcroute Section cleaned down for the visit of an international group of Quaternary researchers in 1986. The pale band at the top is Upper Devensian loesses with some development of limons à doublets. The left dipping beds to the bottom right represent a dune sand developed on an 8 metre O.D. raised beach. This in turn is underlain by earlier loessic layers (by the figure in white).

cuts the earlier 'dolerites'.

East of the Noirmont headland, and about 400m north of the lighthouse, a further deposit of raised beach can be examined a little way down the beach (609467) (Figure 9, location 5). This cobble gravel is up to 0.7m thick and cemented by iron oxides so that it forms an upstanding mass in the modern beach. This beach remnant lies at about mid-tide level and probably illustrates the height range of the 8m beach occasioned by the large tidal range in Jersey. The small bay in which the beach remnant occurs is backed by good sections of head with boulders of granite, set in a matrix of angular feldspar and quartz sand from the weathered granite plus silt particles from the loess. In these head sections, and others north to St Aubin, Palaeolithic flint implements have occasionally been found (Keen, 1978b).

The northern contact of the granite has a lobate form which can be demonstrated by walking along the old railway track, now a public footpath, from St Aubin towards Pont Marquet. Begin just north of the old tunnel (605489): Brioverian sediments are exposed in cuttings for the first 500m when they give place to a lobe of the granite a few metres beyond an old quarry in Brioverian, on the south side of the footpath. The granite continues for about 400m until, just west of another disused quarry, the sediments reappear. The attitude of the sediments is unaffected by the granite, though against it they become tougher and contain biotite flakes. The granite has a fine grained margin that

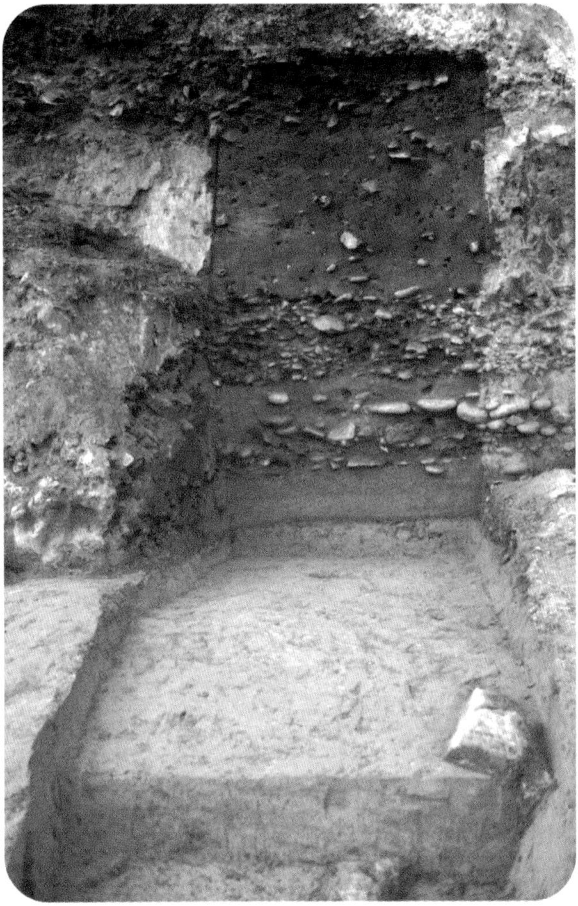

Fig. 11 The lower part of the Belcroute Section cleaned down for the visit of an international group of Quaternary researchers in 1986. Pale loessic layers, exposed in the base of the 1 metre wide trench are overlain by beach sand passing up through sand with cobbles into a dune sand with scattered angular fragments of head.

extends up to a metre from the contact. Further lobes of granite are crossed towards Pont Marquet (599490).

Portelet.

The cove of Portelet is backed by excellent sections in late Middle and Late Pleistocene sediments. To the east of the bay sandy, gravelly head of the Portelet Member rests on the rock platform cut in granite. The basal layers of this head contain coarse marine sands, the product of a high sea-level event

Fig. 12 La Cotte de St Brelade headland from the south showing the main cleft in which the cave is located and which lies on a fault which extends past the stack in the medium distance and along the gully in the foreground.

(Keen *et al*., 1996). The head can be traced laterally to the centre of the bay (Figure 9, location 6) where it rises beneath a beach composed of granite boulders and finer clasts. This in turn is covered by a dune sand with clayey layers within it interpreted (Keen *et al*. 1996) as fossil soils. The dune sands become progressively more stony upwards and pass into true head. Further west, immediately east of the beach café, this head was cut in the Pleistocene by a gully which subsequently filled with loess. This latter deposit is mostly decalcified, but its lower levels, from *c*. 4m above the usual beach level, are less disturbed and have yielded a sparse land molluscan fauna including the arctic/alpine species *Columella columella* (von Martens) (Rousseau & Keen, 1988). This long sequence, although undated by radiometric means, appears to span two temperate high sea-level phases in OI Stages 7 and 5 and multiphased cold stages in 6, and 4-2.

Ouaisné (low tide needed).

At Ouaisné (595476) the sea wall was built during the German occupation and landward of it is a flat, formerly marshy area backed some distance away by a degraded cliff possibly initially cut in the Middle Pleistocene, though it could be much older. Walk westwards across the beach and across the tumbled boulders from the old quarry seen on the left and thence southwards through a small gap in the foreshore rocks to La Cotte Point (Figure 9, location 7 at

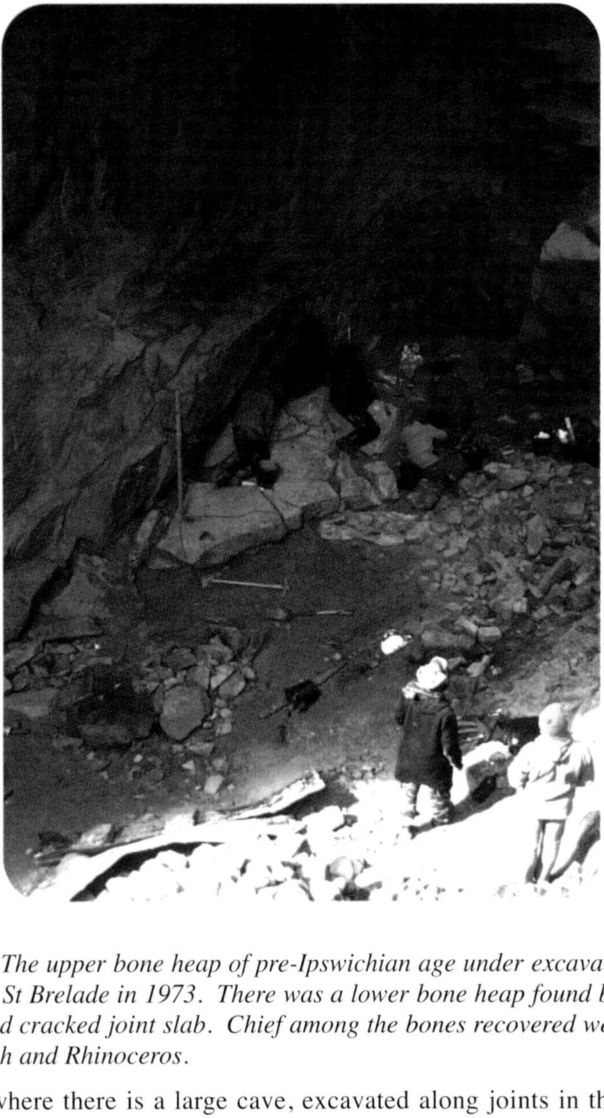

Fig. 13 The upper bone heap of pre-Ipswichian age under excavation in La Cotte de St Brelade in 1973. There was a lower bone heap found beneath the fallen and cracked joint slab. Chief among the bones recovered were those of Mammoth and Rhinoceros.

593475) where there is a large cave, excavated along joints in the porphyritic variety of the Southwest Granite, earlier filled with head (Figure 12). Excavations at the site over the past century have established that occupation of the cave by people using Mousterian cultures extended from at least 300Ka ago to an early stage of the Devensian glacial. As well as many thousands of artefacts of flint, quartz and dolerite, numerous vertebrate remains have been recovered from the cave fill. Most of these are of large cold climate mammals such as *Mammuthus primigenius* (Blumenbach) (Mammoth) and *Coelodonta antiquitatis* (Blumenbach) (Woolly Rhinoceros), but in 1910 nine teeth of *Homo sapiens neanderthalensis* King were found in the loessic head of the upper cultural layers with another four coming to light a year later (McBurney &

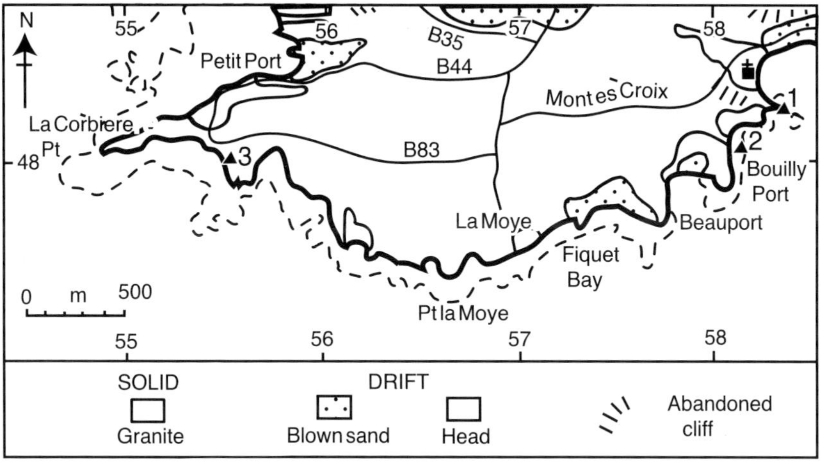

Fig. 14 Geology of south-west Jersey. (IPR/15-29C British Geological Survey. © NERC. All rights reserved).

Callow, 1971; Callow & Cornford, 1986) (Figure 13). **The cave is a protected site and entry is forbidden.**

Two rock types of the Southwest Granite Complex can readily be examined along the above traverse at Ouaisné: the main coarse biotite-hornblende granite in the old quarry with the striking joint surface and the porphyritic granite at and around the point (Pembroke & D'Lemos, 1996)

St Brelade's Bay (west).

On the western side of St Brelade's Bay, from St Brelade's church to Bouilly Port (581481), it is possible to traverse southwards across (1) the coarse granite on the headland south of St Brelade's church where it is cut by basic dykes (Figure 14, location 1), (2) the quite narrow outcrop of porphyritic granite in Bouilly Port (581482) with its 1cm crystals of potassic feldspar set in a finer-grained matrix and (3) the fine-grained aphyric pink granite into which the porphyritic granite rapidly passes. A north-south mica lamprophyre dyke is exposed on the shore (581481), but a careful search is needed to locate it (Figure 14, location 2 at 581481). This fine-grained granite is largely restricted to the area around Beauport (578479) and Fiquet Bay, (575477) but it may also be found on the foreshore South of Ouaisne west of Fliquet (see location 7, Figure 9). The fine-grained granite passes through the porphyritic variety, into the coarse biotite-hornblende granite once again which extends westwards to Corbière (550481) and then northwards to the contact with Brioverian sediments in St. Ouen's Bay. Note that the headland between Bouilly Port and Beauport can only be rounded at low tide, and that the land

Fig. 15 The Les Landes plateau rising from c.6.5 m to c. 80 m above mean sea-level separated from the current shore platform and lowland by the multi-phase fossil cliff line. The fossil cliff line is developed along the contact between the Northwest Granite (plateau and cliff line) and the Jersey Shale Formation (lowland). Neolithic peat horizons lie beneath the beach and behind parts of the German coastal sea wall.

around the point dividing the two is private. At Beauport there is some spectacular castellated granite and other cliff features. Return to St Brelade's

Church via the Mont ès Croix road from Beauport or the footpath from Beauport car park.

La Corbière.

From St Brelade, take the A45 up the hill to the traffic lights at Red Houses, and then follow the main (A13) road westwards turning left along the B83 to Corbière (549481). The murally jointed cliffs are formed of coarse, hornblende-biotite granite of the type examined at Noirmont Point (see page 22); the granite is intruded in this area by a few basic dykes which trend just north of east. The Jersey granites are not richly mineralised, but the remains of the old trial diggings for iron ore are still visible at the head of a prominent gully on a line between the former German ranging tower (now a coastguard station) and Corbière lighthouse (Figure 14, location 3). Nearby there are small dumps, now grassed over, of material thrown out when, in the 19th century, the ore was worked commercially for a time.

St Ouen's Bay.

The Holocene deposits of St Ouen's Bay can be examined in part of either itinerary 4 or 5. The bay can be best seen from view points at La Carrière in the south or L'Étacq in the north. The back of the bay consists of an abandoned cliff with its base at 20-25m above mean sea-level, and topped by a peneplained summit surface at 65m (Keen, 1993) (Figure 15). The current form of the bay is due to late Holocene sand blowing which in the south has carried to the top of the plateau and travelled 2km inland to Pont Marquet. Most of this blown sand is younger than 4000 BP as indicated by radiocarbon dates on underlying peats (Jones *et al.*, 1990). Depending on modern beach levels, usually lowest in spring, the centre of the bay for up to 1km south of Le Port may exhibit a 'submerged forest' of peat and organic sediments up to 3m thick The peat has wood and *in situ* stumps of *Alnus* (alder) and *Corylus* (hazel), nuts of *Corylus*, domestic animal bones, and potsherds, the latter often stained black by the peat.

THE NORTHWEST GRANITE COMPLEX
ITINERARY 5

L'Étacq and Le Pulec (half-tide to low tide).

In the west corner of a disused quarry, landscaped as a small amenity with parking, (558543) on the north side of the B35 road near L'Étacq, the outer coarse granite of the annular Northwest Granite Complex, which contains a little disseminated molybdenite, intrudes northeasterly dipping sediments of the Brioverian Jersey Shale Formation (Figure 16, location 1). Further exposures of these sediments at the roadside opposite the slipway below Grand Étaquerel (548547) (Figure 16, location 2) show sole structures on the undersides of the bedding surfaces. The sediments are best exposed however in the low rocks on either side of the slipway itself. Here, siltstone units up to 1m thick are interbedded with purplish mudstones showing thermal spotting. The rocks dip at about 40° NE and sole structures, flame structures and penecontemporaneous slumping can be observed, particularly in thinly-bedded units in exposures near to the sea-wall and in boulders. Helm & Pickering (1985) have shown that the sedimentary structures here indicate deposition from relatively high-density turbidity currents.

The contact between the Northwest Granite and the Brioverian sediments runs through Le Pulec. The coast northwest of Le Pulec follows closely the line of the contact and the precise trace can best be made out from the small headland southwest of the bay. Descend to the beach by the slipway (Figure 16, location 3). In a gully on the landward side of the slipway, there is an exposure of the 8m raised beach with some cobbles overlain by about 6m or so of head with angular rock fragments becoming more loessic upwards.

The near vertical, stoped contact between the granite and the sediments is exposed for about 400m along the northeastern side of the bay. Numerous aplite and other veins, some of porphyritic microgranite and some composite, cut the sediments. Dolerite dykes also cut the Brioverian sedimentary rocks. Most of the well rounded cobbles in the bay, including those now behind the sea-wall, are derived from angular blocks of granite jettisoned down the cliffs when German fortifications were constructed between 1940 and 1945. At half tide a small sphalerite-galena-dolomite vein is exposed and runs just north of west from the extreme northeast angle of the bay. The vein has a complex mineralogy (Ixer & Stanley, 1980) and was formerly worked for ore but has long been abandoned. **Do not collect from this locality.**

Le Pinacle (half-tide to low tide if sill is to be fully examined).

From Le Pulec either take the cliff path on foot to Le Pinacle (545554) or drive up the hill, noting the view southwards across St Ouen's Bay formed by erosion

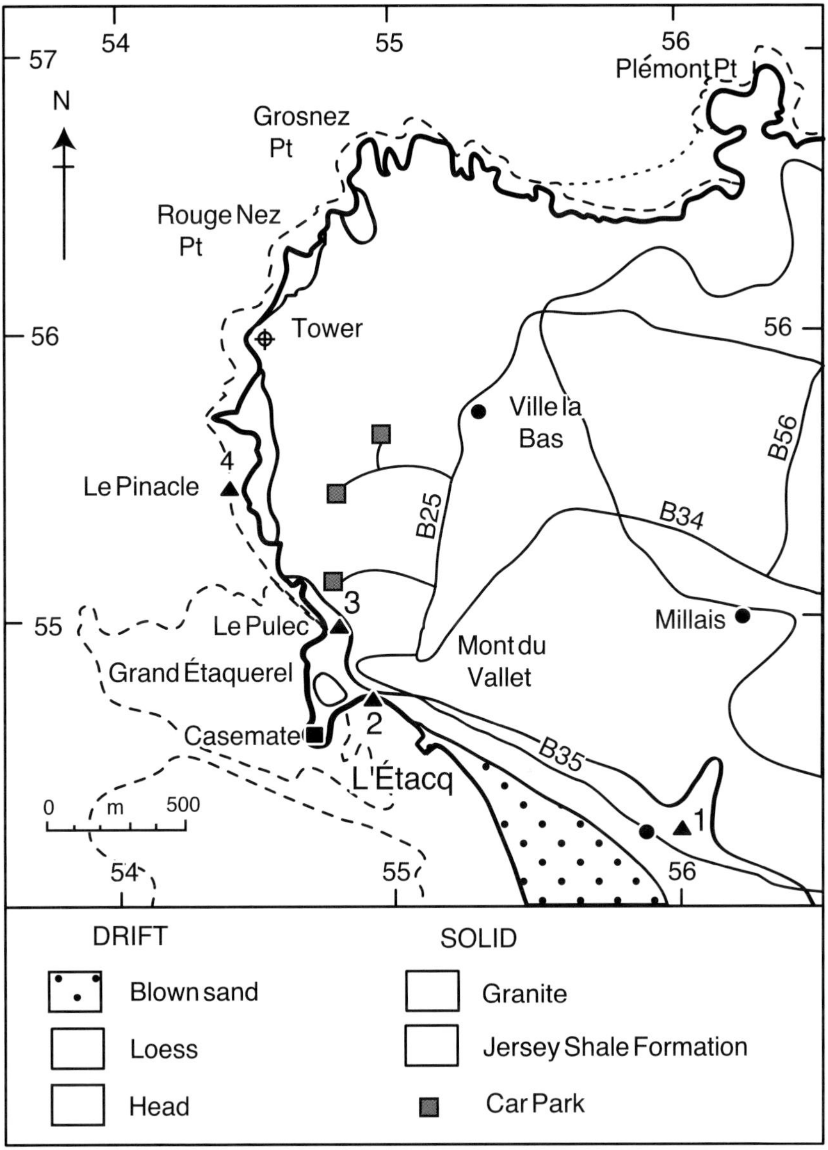

Fig. 16 Geology of north-west Jersey. (IPR/15-29C British Geological Survey. © NERC. All rights reserved).

of Brioverian sediments between the Northwest and Southwest Granites and also the old cliff-line inland. Take the road and tracks to the middle of the three parking areas shown: Le Pinacle is some three hundred metres further on, a promontory connected to the mainland by a narrow isthmus (Figure 16, location 4). This isthmus is penetrated by a north-south natural tunnel.

The steep seaward face of Le Pinacle is caused by the erosion of a basic sill about 1.5m thick which occurs at its base and which dips southeast beneath it. In 1886, the Reverend Ch. Noury, the father of Jersey geology, predicted that the sill under Le Pinacle would erode away causing half the mass to fall within 100 years. It is still there seemingly unchanged. Another sill occurs at a higher level in the cliffs to the south and there are eight such sill-like intrusions between Le Pulec and Grosnez (549567) though most are difficult of access. The isthmus joining Le Pinacle to the mainland is formed by undercut head, which may be examined with caution at its northern end. The eastern side of the gully shows a wave-cut notch marking the 18m raised beach below which, at about present high water mark, there are wave-cut notches representing the 8m sea-level.

Le Pinacle isthmus is also a multi-occupation archaeological site spanning Neolithic to Gallo-Roman times. Many Neolithic stone axes and perforated hammers were manufactured on the spot from dolerite taken from the sill and some were exported to the other islands. The site was used in Gallo-Roman times and the outline can be seen of the rectangular temple built during this period. During the occupation of Le Pinacle, periodic incursions of blown sand from St. Ouen's Bay covered the cultural levels.

Continue northward along the cliff top path past the German radar tower to Rouge Nez Point (546563); further sills in the granite are visible en route from the cliff top.

ITINERARY 6
RONEZ AND SOREL POINT

Granite-diorite-gabbro relationships.

Follow the Route du Nord westwards towards Ronez Point (618571). A good eastward view along the coast may be obtained from the car park (625557) and at low tide the 8m platform is particularly well shown.

Continue past Ronez Quarry to Sorel Point (Figure 17, location 1 at 612571). It is not possible to visit Ronez Quarry, but the workings can be seen from Sorel Point and the relationships between the various rock groups can be observed around Sorel Point itself. In the past much diorite quarried at Ronez was exported for setts and curb stones. More recently the stone has been used as bulk fill in the new harbour areas of St Helier. At present Ronez is the island's main supplier of tarmac and concrete (see figure 33 in Rose *et al.* 2002).

Descend to Sorel Point. **Take care even in calm weather when standing close to the water for there is commonly a swell which can sweep the unwary from the rocks into deep water.**

The nature and origin of the complicated relationships between the acid and more basic rocks at Sorel Point - and elsewhere in the island - have been the subject of

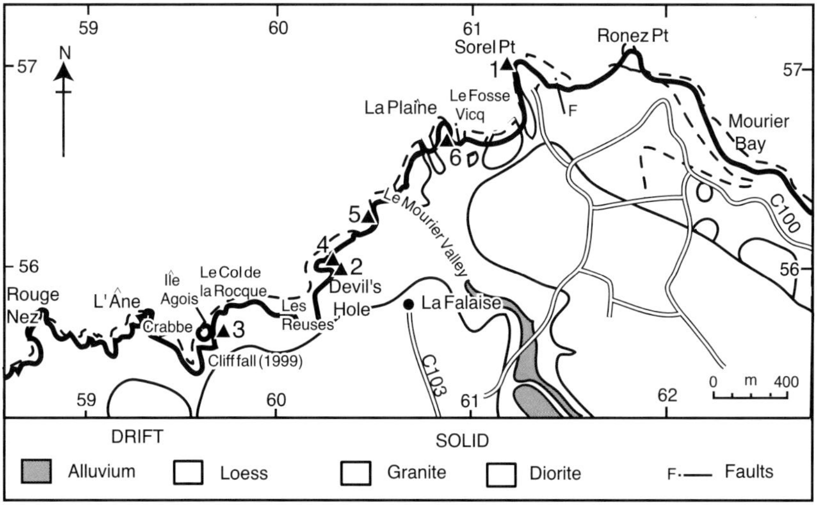

Fig. 17 Geology of part of the north coast of Jersey. (IPR/15-29C British Geological Survey. © NERC. All rights reserved

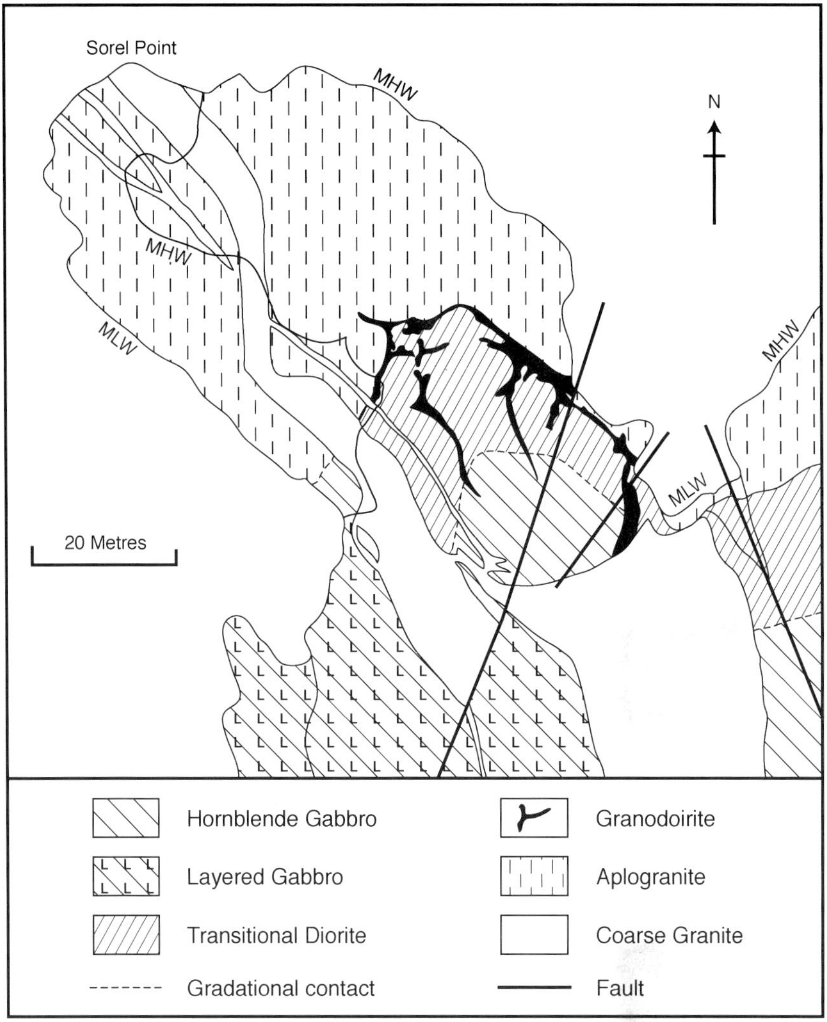

Fig. 18 Geology of Sorel Point (after Salmon, 1998)

much discussion from the first major account by Wells & Wooldridge (1931), through Wells & Bishop (1955), Bishop & Key (1983) to the present time Salmon (1987, 1992, 1996, 1998). The contact relationships and other features were previously interpreted as being due to metasomatism, recrystallisation and rheomorphism of solid gabbro in response to granite intrusion. These interpretations were in line with ideas prevalent at the time. Comparison with other, well documented accounts of magma mingling now shows that the origin of the main features at Sorel Point is the result of physical interaction between coexisting magmas. Infiltration metasomatism, in a magmatic state rather than the classical solid- state form, has produced a number of lithological variants, although, in the main, this process only operated on a small scale.

The tip of Sorel Point, like Ronez Point to the east, is composed of a pink, even-grained aplogranite. There is also a dyke-like body of coarse-grained, grey granite cutting both the aplogranite and all the other rocks present on the headland. This runs more or less north-south. Looking landward from the Point the aplogranite is seen to be separated from the blue-grey diorite almost everywhere by a thin sheeted zone of granodiorite (Figure 19). The details of the aplogranite/granodiorite/diorite relationships are of special interest.

The granodiorite and diorite are mostly in sharp contact, but with lobate forms which are usually crenulate at the millimetric scale. The grain size of the diorite reduces towards the granodiorite and this is interpreted as a chilled margin assumed, in the magma mingling context, to result from the dioritic magma being chilled by the cooler granodioritic one. Rarely, the granodiorite veins the aplogranite; by contrast it penetrates the overlying diorite extensively, in a number of places, as visually striking pale coloured pipes presently plunging some 50°N, indicating post-formational tilting (Salmon, 1996).

Fig. 19 Magmatic interactions at the north end of Sorel Point. Diorite (D) is separated from aplogranite (AG) by an irregular sheet of granodiorite (GD). The diorite has irregular contacts and a fine grained, chilled, margin. The contacts indicate that all three were initially present as coexisting magmas.

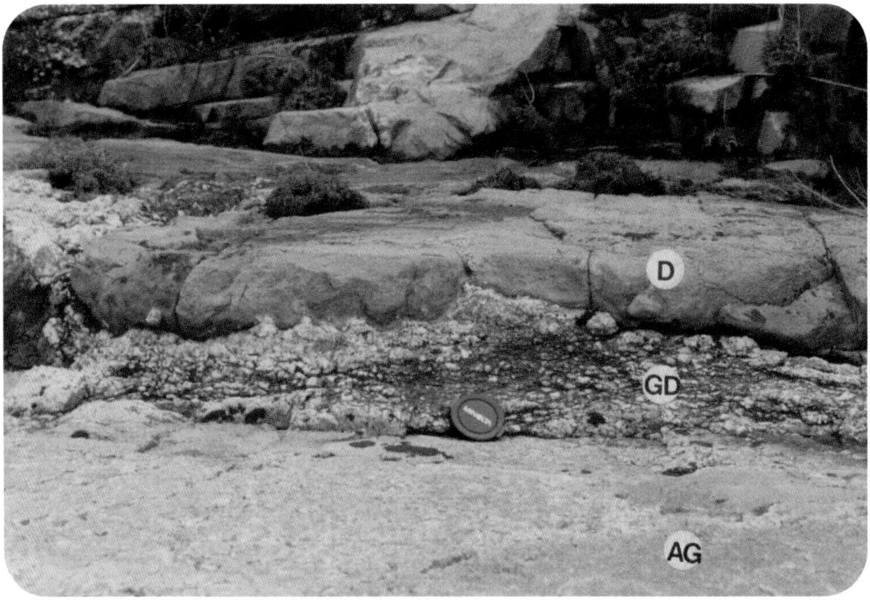

Fig. 20 Contact relationships at Sorel Point. Diorite (D) is separated from aplogranite (AG) by a sheet of granodiorite (GD). The diorite has a crenulate, fine grained (chilled) margin, typical of physical interaction between coexisting magmas.

Please do not collect from these structures – they should be preserved. There are also many enclaves of diorite in the granodiorite and these are of many sizes and forms though usually the contacts are sharp and with finer-grained, crenulate margins similar to the main diorite contact (Figure 20).

As the diorite is followed uphill away from the contact zone it becomes coarser-grained and is characterised by ocelli of quartz or quartz and pink potassic feldspar about 2-5mm across rimmed by small amphibole crystals, and also by patches of quartz and potassic feldspar. Salmon (1992, 1998) has interpreted the diorite (which is referred to in Figure 18 as Transitional Diorite) as having been produced from gabbroic magma by fluid infiltration from the underlying granitic rocks.

Continuing uphill the ocellar diorite gives way gradationally to hornblende gabbro with a poikilitic texture in which amphibole crystals up to 4mm across, often with cores of clinopyroxene, enclose laths of grey-white plagioclase. Sometimes relic olivine is visible in thin-section. Continuing back up the descent path, further gabbros are encountered. These form part of a series of layered hornblende gabbros which dip c.45°S.

Salmon (1992, 1998) has shown that the headland comprises four separate intrusive episodes, each consisting of at least two (usually more) lithologies which were present as coexisting magmas. The intrusive episodes were separated from each other by an interval sufficient for the earlier group of rocks to crystallize completely, giving rise to sharp, angular (i.e. brittle) contacts between episodes. Further intrusive episodes can be identified within Ronez Quarry. The rocks described above at the northern tip of the Point form part of the earliest of the intrusive groups. The coarse granite seen at the northern tip forms part of the latest intrusive group and may be an offshoot of the main Northwest Granite.

Access to the tidal rocks between Sorel Point and Le Fossé Vicq is dangerous and should only be undertaken by those with experience of both tides and cliffs.

ITINERARY 7a
DEVIL'S HOLE TO L' ÂNE AND GRÈVE DE LECQ

The Devil's Hole

The Devil's Hole - *Creux de Vis* in the vernacular patois - is reached by following the main road westwards from St John's Church (B33) and, after about a kilometre, turning right along the C104 and turning right again northwards at the second crossroads (C103). There is a car park at the Priory Inn (607558). Take the signposted path down the valley to the Devil's Hole (Figure 17, location 2) from the inn parking: when the path opens out to the sea there are excellent views to the Paternosters reef and beyond, and to the west, to Sark and Guernsey.

The creux, or hole, first comes into view to the south or left of the path by a protective railing (603559). This deep and wide hole has developed by the collapse and enlargement of its southern end probably the result of its position on a west trending crush zone. The resulting marine erosion and continued collapse has produced the cliff-bound 'beach' seen now. The crushed and broken granite on the eastern wall of the creux still regularly falls. The beach is at present connected to the sea by a through cave running northward underneath the path. Follow the path seaward and left along and around the small conical hill (the rocks here are coarse granite cut by occasional veins of aplite) to reach the viewing platform.

A small mica lamprophyre dyke was recorded on the eastern side of the creux in a small cave, when it was possible to descend by steps. This dyke lies off the trend of the main dyke. The main dyke on the eastern side of the creux is covered by scree, but boulders of lamprophyre are usually recognisable among the rocks on the beach below. The lamprophyres here have achieved a sort of fame since it is the type locality for the lamprophyre jerseyite. The approximately north-south trending cave can be seen at the northern end of the creux at most stages of the tide.

Retrace your steps along the same path, noting on the way the hanging valley to the north of the path which carries a small stream that falls some 15m to the sea. Having regained the inn car park, there are two cliff top paths, one northeastwards to Sorel Point, the other westwards to l'Âne and Grève de Lecq. Take the Grève de Lecq path which soon opens out on to the edge of fields above the Devil's Hole and Les Reuses (602557). A good view of the Devil's Hole is obtained from the cliff path on the western side of Les Reuses, and of a basic dyke intruded into the granite seaward of the conical hill. From the promontory of Le Col de la Rocque (598558) there are good views eastwards to the dioritic rocks at La Plaine (608567) and Sorel Point, and westwards along the granitic rocks to Plémont Point (563569). At half tide or below a shore platform is visible at Les Reuses cut in granite at the 8m level, just about the present high water mark.

Follow the path westwards to Île Agois (Figure 17, location 3), a stack which can be reached only by walking at low tide from Crabbé. Île Agois is an important occupation site with the foundations of about 20 huts excavated into its comparatively gentle west-facing slopes. Roman and Dark Ages coins have been found indicating an age for the occupation in Dark Age times.

The path descending to Crabbé (596555) is presently closed and should not be attempted. As a substitute, continue WNW along the path towards l'Âne and look back down on Île Agois with its caves and gullies. A significant rock fall (1999) from the north-south trending cliff of the Île Agois can be seen partially filling the 'beach' area ESE of the island. It will prove interesting to see how long it will take to be broken up and removed by the sea. Follow the cliff path to l'Âne and thence inland past the old Composting Site and Crabbé Farm. To visit Grève de Lecq, turn westwards at the road junction a short distance south of Crabbé Farm and follow the road to the bay.

There are good exposures of the porphyritic granodiorite in the Northwest Granite around the beach. At the eastern side of the bay, there is a tough north-south trending mica lamprophyre dyke, a little over 1m wide containing small felsic inclusions in places. The granite to the east of the dyke is sheared for about 1.5m, and east-west faults must be inferred to account for the way in which the dyke is offset as it is traced southwards. Further dykes are exposed at the western side of the bay: a mica lamprophyre occurs behind the German bunker and to the north of the Casino cafe and a thin basic dyke used to be exposed a short distance north of this.

ITINERARY 7b
DEVIL'S HOLE TO SOREL POINT

The cliff path to Sorel Point begins by the car park behind the Priory Inn. Begin by walking seawards along the heights above the line of the small valley to the Devil's Hole; a good view of the creux can be obtained by looking southwestwards as the path approaches the coast (Figure 17, location 4). On reaching the cliffs, turn northeastwards and take the narrow path closest to the cliffs that leads downhill and to the east towards Le Mourier Valley. Cross the stream in the valley by the wooden footbridge (607562) and follow the valley path seawards along the northern side of the stream, noting that the valley is partly filled with head.

The stream eventually falls some 10m into the sea at the head of a narrow inlet that follows a small fault in the granite (Figure 17, location 5). The sea has cut back across the valley and isolated the stream from its continuation across the inlet. Follow along the ledge on the eastern side of the gully; the granite here is coarse-grained and contains a few dark inclusions, some of which contain feldspars similar to those in the granite. Note also the aplite veins, up to 20cm in width, which dip gently southwards.

Return up the valley to the bridge and take the path that climbs out of the valley northwards to the heights. Follow this path across outcrops of granite to the point, by a hairpin bend, at which the first northwestward sea view is obtained. The rocks exposed at sea-level on the headland now visible are diorite veined by granite, and just above high water mark there are five 'pipes' of pegmatitic granite in the diorite, each about 0.5m across; one of these has a dark rim and is inclined steeply southeastwards. Follow the path northeastwards noting an exposure of granite in an old quarry with a seat, by the path. Diorite veined by granodiorite is next exposed along the path as it rises to the headland referred to above (608565). From the top of this headland, and by a seat where the path swings round to turn inland, there is a good view eastwards to La Plaine (608567) and beyond to Sorel Point. Both show either diorite veined by grey to pinkish grey granite or granite containing abundant diorite inclusions. With care it is possible to descend the sharp ridge of La Plaine (Figure 17, location 6) towards the sea to examine more closely the relationships between diorite and granite. The deep cave immediately to the east of La Plaine contains a basic dyke. The cliff path continues around Le Fossé Vicq to Sorel Point which can, of course, form the starting point of the walk.

NORTHEAST JERSEY
ITINERARY 8

Côtil Point.

Take the C101 road north from St John's Church (627554) and turn right at the junction with the Route du Nord. Several old quarries near the road expose coarse-grained Northwest Granite (Figure 21, location 1). A wide track descends seawards at the point where the main road turns inland up a small valley. The upper part of the split track leads, as it bears right around the hill, to the working quarry of La Saline (Figure 21, location 2). Permission should be sought of the owners for access, though the workings are of greatest interest for the modern diamond cutting machinery operating on imported stone. The olivine gabbro within the quarry

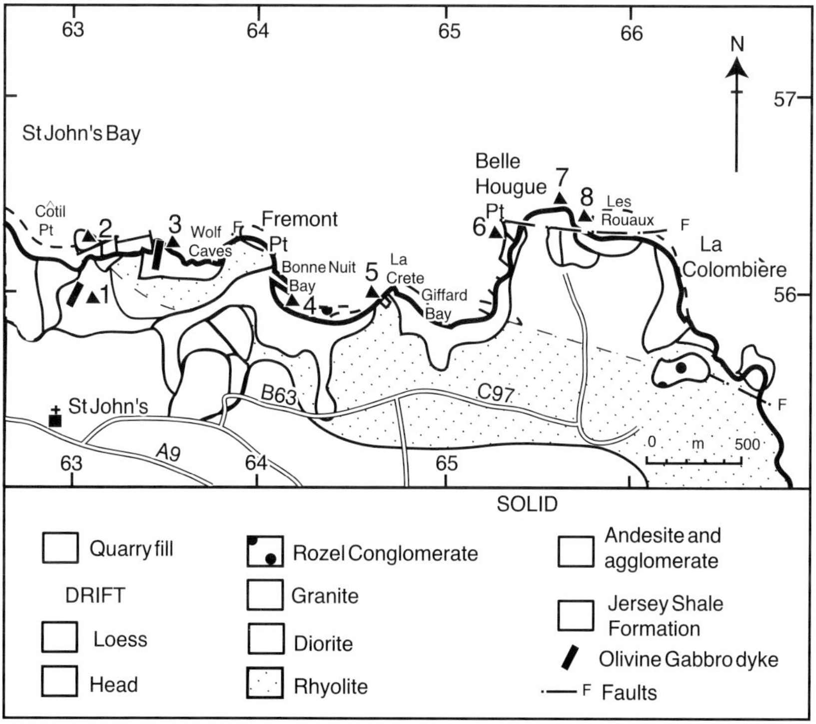

Fig. 21 Geology of the area around Belle Hougue Point. (IPR/15-29C British Geological Survey. © NERC. All rights reserved).

(probably the faulted continuation of the Wolf Caves gabbro some 350m to the east (Figure 21, location 3)) is buried completely beneath a quarry road.

The aplogranite of part of the quarry was formerly worked at Mont Mado (637556) and was much used as a building and ornamental stone with records of its use in both the Neolithic monument and medieval chapels at La Hougue Bie: the quarries of Mont Mado have been filled with waste and are no longer visible.

Return to just short of the exit on to the road and take the other track downhill beyond the signposted cliff path, past the first vehicle barrier to the box-like hut at the hairpin bend. A fisherman's path descends directly to the shore from behind the hut. Follow the path on to the shore platform cut in granite at the 8m level. Walk seawards across a fault on to pinkish grey silicified rhyolite which is intruded by aplogranite similar that at Sorel Point and Ronez Point (Figure 21, location 2). There are many tongues and veins of aplogranite in the rhyolite and careful examination of joint surfaces will reveal a coating of muscovite flakes formed as a result of potassium metasomatism. The rhyolite also contains foliated basic dykes which are cut by veins of granite.

Moving eastwards a short distance along the shore platform note the wave-cut notch and perched sea caves. At low tide, descend to the beach where the coast turns sharply southwards. At the base of the gullies here there are exposures of andesite intruded by aplogranite. Andesite continues from this point eastwards until it gives place within a few hundred metres to rhyolite. **This part of the coast is very difficult of access and can easily be cut off by the rising tide. Return to the cliff top by the same route.**

Bonne Nuit Bay, Giffard Bay.

Low tide is essential for the examination of the volcanic rocks of Giffard Bay, and low spring tides are best: great care should be taken to avoid being cut off by the incoming tide. Most of the rocks exposed at low tide in Bonne Nuit Bay (644559) are of cream, silicified rhyolitic ash showing little of its original structure. A search of the first exposures met after crossing the pebble beach from Bonne Nuit harbour (Figure 21, location 4) will reveal a Pleistocene beach deposit at mid-tide levels containing small pebbles up to 5cm long, set in a sandy ferruginous matrix. Like all the raised beaches on Jersey's north coast the Bonne Nuit raised beach contains quantities of flint, in this exposure up to 5 per cent of the total stones, probably derived from submarine outcrops of Chalk to the north and west of the island. The beach lies just above mean sea-level (*cf.* Noirmont) and forms a further example of the 'submerged beach' of Mourant (1933) and probably shows the former height range of the 8m beach. Follow the rock platform eastwards noting the cliffs in thick undifferentiated head with beach pebbles in its lower layers. Just west of La Crête Point (647560) there is a prominent layer of andesite about 10m thick downfaulted to the west (Figure 21, location 5). It is succeeded by ignimbrites at the point itself, where the succession dips to the east.

Giffard Bay lies east of La Crête Point and is best reached by following the small road leading eastwards from the C98 road passing behind the Cheval Rock hotel/apartments. Take the path downward to the fort at La Crête and descend to the beach on the eastern side of the fort. The western part of Giffard Bay is composed of a series of coarse to fine-grained acid tuffs with some agglomerates. The succession here dips to the west.

The dissected rock platform in Giffard Bay is an old 8m level backed by high cliffs of head and loess. Recent erosion has carved deep gullies into the 8m platform, but a wave-cut recent beach platform has been eroded and this, when clear of cobbles, affords excellent sections in pyroclastic rocks. Similar wave-polished sections are exposed in the sides of the gullies in the 8m platform where in particular there are excellent examples of purple and green air-fall tuffs and agglomerates. At low spring tides the rough scramble over the rocks can be avoided by walking along the sandy beach seaward of the rock platform.

In the southeast corner of Giffard Bay there is a fault-bounded exposure of grey, laminated, Brioverian mudstones convoluted by penecontemporaneous slumping. North of this there occurs l'Homme Mort Conglomerate which contains rounded pebbles of andesite, granite, and Brioverian sediments, together with some jasper. The conglomerate continues northwards to Long Echet (653563) (Figure 21, location 6), and most probably represents a sedimentary interlude during the period of andesitic volcanism. Care should be taken on this section to avoid being cut off by the tide. Return to the road by the same route. For a map and description of this area see Roach *et al.* (1986, fig. 33) which also has the older, rather dated but still useful map by Casimir & Henson (1955, fig. 34).

Les Rouaux (half tide).

Les Rouaux (658563) is the small bay east of Belle Hougue Point which is reached by following the road leading north from the C97 road east of the BBC transmitter station at Les Platons. Follow the path to the easterly and lower of the two headlands at Belle Hougue Point (Figure 21, location 7). This also provides a good viewing point both westwards and eastwards (Figure 22).

Between the two headlands to the west of the viewing point there is a complex of caves, two of which - Belle Hougue Caves I and II - contain raised beaches cemented by travertine. The beaches occur in the backs of the caves at heights at about the 8m level. Nine species of marine molluscs have been obtained from the beach gravel in the main cave — Belle Hougue I — and indicate a sea temperature up to 3°C warmer than that of the present English Channel. Bones of an insular dwarfed race of deer, *Cervus elaphus jerseyensis* Zeuner have also been found. Lister (1995) concludes that isolation from the mainland allowed dwarfing to one half the size of modern deer in as little as 10,000 years. U-series dates from the travertine cement of the beach confirm an age which correlates with OI Sub-stage 5e worldwide (Keen *et al.*, 1981) and thus with the Ipswichian/Eemian interglacial. The upper cave contains a fine wave-smoothed notch, and a smaller remnant of the 18m raised beach also cemented with travertine. **Access to these caves is difficult and hazardous and should not be attempted.**

Follow the old cart track, now part of the cliff path, downwards and eastwards. Just east of a wet patch of ground crossed by a stream, a steep, grassy cliff

Fig. 22 The west to east running gully dominating the view at Les Rouaux by Belle Hougue Point separates granitic rocks to the north (left) from Brioverian Shale Formation sediments (bottom right) which pass up into contemporaneous andesitic arc volcanic rocks away from the observer.

path descends to Les Rouaux and a shore platform at the 8m level with a good 2m high wave-cut notch at and to the west of the point where the path reaches the shore (Figure 21, location 8). The shore platform is cut in Brioverian sediments showing penecontemporaneous slumping and fracturing and which have complicated fold patterns with steeply plunging fold axes most probably due to movement of the east-west fault which separates the sediments from the igneous rocks to the north. Cross the gully at the western end of the bay to the plutonic rocks which lie to the north of the fault. The first rocks to be met are pink granite and syenite, but these give place northwards, and beyond a small fault, to rather altered diorites. The granitic rocks vein and invade the diorite, and in places are cut by basic dykes. At the eastern end of the shore platform the Brioverian sediments give place upwards to andesitic pyroclastic rocks just beyond a shingle embayment. Return to the viewing point *en route* to the road and look back eastwards across Les Rouaux. Notice that, at low tide, the gully marking the line of the fault can be traced right across the bay to the foot of La Colombière Point.

ITINERARY 9
BOULEY BAY
(High tide to half tide)

Bouley Bay.

Descend to Bouley Bay by the C102 road. Around and to the north of a small jetty (670548) there are exposures of green and purple silicified rhyolitic ignimbrites which have long been known for the spherulites they contain. Good small spherulites can be examined in the rocks at Porteret near to the sea wall by the road leading northwards to the jetty.

The spherulites in the Jersey Volcanic Formation vary greatly in size from about the size of a pin-head to 30 centimetres or more in diameter. It is best to collect specimens from the beach pebbles; a search will soon reveal a variety of size and form of spherulites better than those in the rocks exposed immediately around (Mourant, 1932, 1933).

La Tête des Hougues (half tide).

At the foot of La Tête des Hougues is a small unnamed cove (679544) which forms the southeast corner of Bouley Bay. It is here that the junction between the Jersey rhyolites and the overlying Rozel Conglomerate Formation is exposed.

The Bay is reached by following a path leading northwards from the C93 road about 300m northeast of Pot du Rocher (677539). A track leaves the road immediately east of a house. Follow the track and path until it joins the open cliff path. Follow this eastwards past the house in the small quarry on the left before descending on the far side. Some ten to twenty metres before the stream a rough path continues the line of the descending cliff path. Follow this path through the bracken ending with a steep scramble into the little cove at the bottom. Take care as the final rocky descent to the cove shingle is very slippery.

The western part of the cove is formed of grey-green rhyolitic ash which dips east. The ash gives place upwards first to angular rhyolitic rubble then to progressively finer sediments. At the foot of the slope marking the western side of the bay there are several metres of red mudstones with occasional coarser, gritty layers. Cornes (1933) recorded fossil rain-prints in these mudstones which in turn give place upwards to the coarse conglomerate characteristic of the bulk of the Rozel Formation. A full account of the alluvial fan deposition at this locality is given by Went & Andrews (1990).

The pebbles in the conglomerate are partly of local derivation and include sub-angular fragments of indurated grey siltstones of Brioverian aspect, and frag-

Fig. 23 Dr Deryck Laming pointing out imbrication in clasts within the Rozel Conglomerate Formation in the cove below the Tête des Hougues. The direction of movement of the flash flood was from the north (left) to the south.

ments of volcanic rocks. Other clasts are of more doubtful origin and include large rounded cobbles and boulders of 'granite' with greenish plagioclases, pebbles of micrographic granite and occasional pebbles of conglomerate (Figure 23). The Rozel Conglomerate Formation is roughly bedded and dips northeast at about 30°. The conglomerate is a late Cadomian alluvial fan conglomerate of late Cambrian or early Ordovician date. Return to the road by the same route or by continuing along the easterly cliff path across the stream.

WET DAYS

On wet days a visit can be paid to the Jersey Museum, at the Weighbridge, St. Helier, or to the Museum at La Hougue Bie, St Saviour (683504). Apart from the intrinsic interest of the La Hougue Bie site, the geological and archaeological collections are kept and displayed here. Maps, guides and other publications may be purchased at both Museums. The Société Jersiaise library in the Jersey Museum complex contains most geological publications of island relevance though with an emphasis on the older ones. Purchases of likely interest and value are (1) the British Geological Survey, Jersey (Channel Islands Sheet 2), 1982 and (2) Bishop, A.C. & G. Bisson, *Classical areas of British geology: Jersey: description of 1:25000 Channel Islands Sheet 2*. London: Her Majesty's Stationery Office for British Geological Survey. 1989. ISBN 0 11 884458 X

ACKNOWLEDGEMENTS

In compiling this guide we have been able to draw on the help of many friends and colleagues. Some of the geological information was obtained during a geological survey of the island as part of a joint project by the Institute of Geological Sciences (now British Geological Survey) (1972-75) (N.E.R.C.) and the States of Jersey and of Guernsey to produce the 1:25000 geological maps of the Channel Islands.
We wish to thank Dr G. M. Thomas and the late Dr A. E. Mourant FRS for their help and advice in the preparation of the original 1981 text. Dr D. Went's work provides the basis for the description and interpretation of the Rozel Conglomerate Formation. Apart from Figure 18 (Dr S. Salmon), the sketch maps are published by kind permission of the Director of the Institute of Geological Sciences (N.E.R.C.) under copyright permit No IPR/15-29C. The typescript was prepared for publication by Ms Gillian West, and the maps were redrawn from the first edition of the guide by Mark Rye and Joanne Beverley (Coventry University). Financial support for the production of the guide came from the Curry Fund of the Geologists' Association.

FURTHER READING

ADAMS, C.J.D. 1976. Geochronology of the Channel Islands and adjacent French mainland. *Journal of the Geological Society, London*, **132**, 233-250.
BATES, M.E., PARFITT, S.A. & ROBERTS, M.B. 1997. The chronology, palaeoecology and archaeological significance of the Marine Quaternary record of the West Sussex Coastal Plain, Southern England, UK. *Quaternary Science Reviews,* **16,** 1227-1252.
BISHOP, A.C. 1964a. The petrogenesis of hornblende-mica lamprophyre dykes at South Hill, Jersey, C.I. *Geological Magazine,* **101**, 302-313.
— 1964b. The La Collette Sill, St. Helier, Jersey, C.I. *Annual Bulletin of La Société Jersiaise*, **18,** 418-428.
— & BISSON, G. 1989. *Classical areas of British geology : Jersey : description of 1:25000 Channel Islands Sheet 2*. London : Her Majesty's Stationery Office for British Geological Survey, 126pp.
— & KEY, C. 1983. Nature and origin of layering in diorites of S.E. Jersey, Channel Islands. *Journal of the Geological Society, London,* **140**(6), 921-937.
— & KEY, C. 1984. Discussion on the nature and origin of layering in the diorites of SE Jersey, Channel Islands. *Journal of the Geological Society. London,* **141**(3), 596-598.
— ROACH, R.A. & ADAMS, C.J.D. 1975. Precambrian rocks within the Hercynides. In: Harris, A.L. *et al.* (editors) *A correlation of the Precambrian rocks in the British Isles,* Geological Society of London Special Report, 6, 135pp.
CALLOW, P. & CORNFORD, J. (editor). 1986. *La Cotte de St Brelade 1961-1978: Excavations by C.B.M. McBurney*. Geo Books, Norwich.
CASIMIR, M. & HENSON, F.A. 1949. Dyke phenomena of the Dicq rock, Jersey, C.I. *Geological Magazine,* 86, 117-122.
— & HENSON, F.A. 1955. The volcanic and associated rocks of Giffard Bay, Jersey, Channel Islands. *Proceedings of the Geologists' Association,* **6**, 30-50.
COOPE, G.R., JONES, R.L. & KEEN, D.H. 1980. The palaeoecology and age of peat at Fliquet Bay, Jersey, Channel Islands. *Journal of Biogeography,* **7,** 187-195.
— JONES, R.L., KEEN D.H. & WATON, P.V. 1985. The flora and fauna of Late Pleistocene deposits at St Aubin's Bay, Jersey. *Proceedings of the Geologists' Association,* **96**(4), 315-323.
CORNES, H.W. 1933. The Jersey conglomerate. *Bullétin Annuel de La Société Jersiaise,* **12**, 118-151.
DALLMEYER, R.D., D'LEMOS, R.S. & STRACHAN, R.A. 1992. Timing of post-tectonic Cadomian magmatism on Guernsey, Channel Islands: evidence from 40Ar/39Ar mineral ages. *Journal of the Geological Society, London,* **149**, 135-147.
— D'LEMOS, R.S. & STRACHAN, R.A. 1994. Timing of Cadomian and Variscan tectonothermal activity, La Hague and Alderney, North Armorican mas-

sif: evidence from 40Ar/39Ar mineral ages. *Geological Journal,* **29**, 29-44.
— D'LEMOS, R.S., STRACHAN, R.A., & MUELLER, P.A. 1991. Tectonothermal chronology of early Cadomian arc development in Guernsey and Sark, Channel Islands. *Journal of the Geological Society, London,* **148,** 691-702.
DAVIES, K.H. & KEEN, D.H., 1985. The age of Pleistocene marine deposits at Portland, Dorset. *Proceedings of the Geologists' Association,* **96**(3), 217-225.
D'LEMOS, R.S., STRACHAN, R.A. & TOPLEY, C.G. (editors). 1990. *The Cadomian Orogeny*. Geological Society of London Special Publication, 51, 423 pp.
HELM, D.G. & PICKERING, K.T. 1985. The Jersey Shale Formation: A late Precambrian deep water siliclastic system, Jersey, Channel Islands. *Sedimentary Geology,* **43**, 43-66.
IXER, R.A. & STANLEY, C.J. 1980. Mineralization at Le Pulec, Jersey, Channel Islands. *Mineralogical Magazine,* **43**, 1025-1029.
JONES, R.L., KEEN, D.H., BIRNIE, J.F. & WATON, P.V. 1990. *Past landscapes of Jersey : Environmental changes during the last ten thousand years*. Société Jersiaise, Jersey. 145 pp.
KEEN, D.H. 1978a. *The Pleistocene deposits of the Channel Islands*. Report of the Institute of Geological Sciences, 78/26 14pp.
— 1978b. A flint flake from Noirmont Point. *Annual Bulletin of La Société Jersiaise,* **22**, 205-208.
— 1981 *The Holocene deposits of the Channel Islands*. Report of the Institute of Geological Sciences, 81/10.
— (ed.) 1993. *The Quaternary of Jersey - Field guide*. Quaternary Research Association, Cambridge. 192pp.
— 1995. Raised beaches and sea-levels of the English Channel in the Middle and Late Pleistocene: problems of interpretation, and implications for the isolation of the British Isles. In: Preece, R.C. (editor) *Island Britain: A Quaternary Perspective*. Geological Society of London Special Publication, 96, 63-74.
— HARMON, R.S. & ANDREWS, J.T. 1981. U series and amino-acid dates from Jersey. *Nature, London,* **289**, 162-164.
— van VLIET-LANOË, B. & LAUTRIDOU, J.-P. 1996. Two long sedimentary records from Jersey, Channel Islands: stratigraphic and pedologic evidence for environmental change during the last 200k yr. *Quaternaire,* **7**(1), 3-13.
LEUTWEIN, F., POWER, G., ROACH, R.A. & SONET, J. 1973. Quelques résultats géochronologiques obtenus sur les roches d'âge précambrien du Cotentin. *Comptes-Rendus de l'Académie des Sciences, Paris,* **276**, 2121-2124.
LISTER, A.M. 1995. Sea-levels and the evolution of island endemics: the dwarf red deer of Jersey. In: Preece, R.C. (editor) *Island Britain: A Quaternary Perspective*. Geological Society of London Special Publication 96, 151-172.
McBURNEY, C.B.M. & CALLOW, P. 1971. The Cambridge excavations at La Cotte de St. Brelade, Jersey — a preliminary report. *Proceedings of the Prehistoric Society,* **37**, 167-207.
MILLER, B.V., SAMSON, S.D. & D'LEMOS, R.S. 2001. U-Pb geochronological constraints on the timing of plutonism, volcanism, and sedimentation, Jersey, Channel Islands, UK. *Journal of the Geological Society, London,* **158**, 243-252.

MONNIER, J.-L., HALLÉGOUËT, B., van VLIET-LANOË, B., MOLINES, N., LAURENT, M., GEIGL, E.M. & HINGUANT, S. 1997. The Finistère : Audierne Bay, Menez Dregan and Gwendrez : 97-101. In : Van Vliet-Lanoë, B., Hallegouët, B. & Monnier, J.-L. (editors): *The Quaternary of Brittany : Guide book of the excursion of the Quaternary Research Association in Brittany*, 12-15 September 1997. Travaux Laboratoire Anthropologie, Université de Rennes, Volume Spécial, Rennes. 132 pp.
MOTTERSHEAD, P.N., GILBERTSON, D.D. & KEEN, D.H. 1987. The Pleistocene marine deposits of Torbay, a re-appraisal. *Proceedings of the Geologists' Association,* **98**(3), 241-257.
MOURANT, A.E. 1932. The spherulitic rhyolites of Jersey. *Mineralogical Magazine,* **23**, 227-238.
— 1933. The raised beaches and other terraces of the Channel Islands. *Geological Magazine,* **70**, 58-66.
NOURY, Ch. 1886. *Géologie de Jersey*. Savy, Paris and Le Feuvre, Jersey. 173pp.
PATTON, M. 2002. The cist grave cemetary of La Motte (Green Island), Jersey: Prehistoric or Medieval? *Annual Bulletin Société Jersiaise*, **28**(2), 252-260
PEMBROKE, J.W. & D'LEMOS, R.S. 1996. Mixing between granite magmas : Evidence from the South-west Granite Complex of Jersey. *Proceedings of the Ussher Society*, **9**(1), 105-113.
RENOUF, J.T. 1993. Solid geology and tectonic background : 1 - 11. In : Keen, D.H. (editor) 1993. *The Quaternary of Jersey - Field guide*. Quaternary Research Association, Cambridge. 192 pp.
ROACH, R., ADAMS, C., BROWN, M., POWER, G. & RYAN, P. 1972. The Precambrian stratigraphy of the Armorican Massif, N. W. France. *Proceedings of the 24th International Geological Congress, Montreal*. Section **1**, 246-252.
— TOPLEY, C., BROWN, M. & SHUFFLEBOTHAM, M. 1986. Brioverian volcanism and Cadomian plutonism in the northern part of the Armorican Massif. Pre-conference excursion guide: *Geochemistry and Mineralisation of Proterozoic Volcanic Suites*. Geological Society of London Special Meeting.
ROSE, E.P.F., GINNS, W.M. & RENOUF, J.T. 2002. Fortification of Island Terrain: Second World War German Military Engineering on the Channel Island of Jersey, a Classic Area of British Geology. In Doyle, P. & Bennett, M.R. (editors) *Fields of Battle* Kluwer Academic, 265-309.
ROUSSEAU, D.-D. & KEEN, D.H. 1988. Malacological records from the Upper Pleistocene at Portelet (Jersey, Channel Islands): comparisons with western and central Europe. *Boreas,* **18**, 61-66.
SALMON, S. 1987. Some relationships within the igneous complex at Sorel Point, Jersey: metasomatism or magma-magma interaction? *Proceedings of the Ussher Society*, **6**, 510-515.
— 1992. *Contemporaneous acid-basic plutonism at Sorel Point, Jersey, Channel Islands*. Unpublished PhD thesis, CNAA/Oxford Polytechnic.
— 1996. Cylindrical granodiorite pipes in the Sorel Point Igneous Complex, Jersey, Channel Islands. *Proceedings of the Ussher Society,* **9**(1), 114-120.
— 1998. The plutonic igneous complex at Sorel Point, Jersey, Channel Islands: a

high-level multi-magma assemblage. *Geological Journal*, **33**, 17-35.
SAMSON, S.D. & D'LEMOS, R.S. 1998. U-Pb geochronology and Sm-Nd isotopic composition of Proterozoic gneisses, Channel Islands, UK. *Journal of the Geological Society, London,* **155**, 609-618.
— & D'LEMOS, R.S. 1999. A precise late Neoproterozoic U-Pb zircon age for the syntectonic Perelle quartz diorite, Guernsey, Channel Islands, UK. *Journal of the Geological Society, London*, **156**, 47-54.
SHORTLAND, R.A., SALMON, S., ROWBOTHAM, G. & REGAN, P.F. 1996. Coexisting acid and basic magmas of the Elizabeth Castle Igneous Complex, Jersey, Channel Islands. *Procedings of the Ussher Society,* **9**(1), 121-126.
SMITH, H.G. 1936. The South Hill lamprophyre, Jersey. *Geological Magazine,* **73**, 87-91.
SOMME, J., PAEPE, R. & LAUTRIDOU, J.P. 1980. Principes, méthodes et système de la stratigraphie du Quaternaire dans le nord-ouest de la France et la Belgique. *Bulletin de l'Association Française pour l'Étude du Quaternaire,* **I** (NS) suppl., 148-162.
TOPLEY, C. & BROWN, M. 1984. Discussion on the nature and origin of layering in the diorites of SE Jersey, Channel Islands. *Journal of the Geological Society, London,* **141**(3), 595-596. Reply by A.C. Bishop & C. Key : 596-598.
WEIBE, R. A. 1993. The Pleasant Bay layered gabbro-diorite, coastal maine: ponding and crystallization of basaltic injections into silicic magma chamber. *Journal of Petrology*, **34**, 461-489.
WELLS, A.K. & BISHOP, A.C. 1955. An appinitic facies associated with certain granites in Jersey, Channel Islands. *Quarterly Journal of the Geological Society, London,* **111**, 143-166.
— & WOOLDRIDGE, S.W. 1931. The rocks groups of Jersey with special reference to intrusive phenomena at Ronez. *Proceedings of the Geologists' Association,* **42**(2), 178-215.
WENT, D. & ANDREWS, M. 1990. Post-Cadomian erosion, deposition and basin development in the Channel Islands and northern Brittany, 293-304. In : D'Lemos, R., Strachan, R.S. & Topley, C.G. (editors), *The Cadomian Orogeny.* Geological Society of London Special Publication, 51.
— ANDREWS, M.J. & WILLIAMS, B.P.J. 1988. Processes of alluvial fan sedimentation, basal Rozel Conglomerate Formation, La Tête des Hougues, Jersey, Channel Islands. *Geological Journal*, **23**, 75-84.

GLOSSARY

Adamellite A variety of granite in which one third to two thirds of the total feldspar is plagioclase.
Amphibolite facies Varieties of metamorphic rocks formed under moderately high pressure and temperature, and carrying hornblende and plagioclase feldspar.
Aphyric Textural term for fine grained non-porphyritic igneous rocks.
Aplite Fine grained, equigranular granitic rocks composed mainly of feldspar and quartz, commonly occurring in veins.
Aplogranite A light coloured equigranular variety of granite carrying small amounts of mica.
Appinite A medium to coarse grained, dark igneous rock carrying conspicuous hornblende in a matrix of plagioclase feldspar.
Autobrecciated A process whereby a previously solidified top of a lava is frag mented by movement within the body of the lava.
Cratonisation A process of stabilisation of the earth's crust .
Crescumulate A layered igneous rock in which elongated crystals are orientated more or less perpendicular to the layering.
Diapir A large igneous mass produced by the vertical movement upwards of a magma column.
Diorite A medium to coarse grained, typically black and white igneous rock consisting of andesine feldspar and common hornblende.
Dolerite A fine to medium grained basic igneous rock consisting of calcic plagioclase feldspars, pyroxenes and iron-ore minerals.
Epidotised The conversion of feldspars into epidote (a complex silicate).
Flame structure Curved injections of shale or siltstone found at the base of sandstones and due to sedimentary loading processes.
Gabbro A coarse grained basic igneous rock formed predominately of calcic plagioclase feldspar and augite.
Granophyre A medium to fine grained variety of porphyritic microgranite with much intergrowth of potash feldspar and quartz in its matrix.
Head A loose deposit of sub-aerially weathered materials capable of being mobilised by processes such as solifluction.
Ignimbrite An indurated volcanic ash in which the fragments and glass shards of the matrix are heat-welded together. Often shows internally a pseudo-flow structure.
Lamprophyre A distinctive group of dark porphyritic igneous rocks occupying dykes, with conspicuous phenocrysts of biotite, hornblende or augite. Minette is one variety rich in biotite.
Limon A French term referring to a mixture of loess and loam.
Loess A homogeneous, unstratified yellowish deposit of silt-sized material,

usually quartz, and regarded as an accumulation of wind-blown dust.

Megacryst A large crystal in a finer matrix or ground mass.

Metasomatism A change in mineral or rock composition brought about by the introduction of material from external sources.

Micrographic texture A fine grained intergrowth of quartz and feldspar resembling hieroglyphic writing.

Ocelli Small circular areas occupied by minerals in igneous rocks and resembling vesicles or amygdales.

Orthoquartzite Silica-cemented quartz-rich sandstone.

Pegmatite A coarse grained igneous rock commonly occurring as veins and dykes cutting through its parent igneous body.

Periglacial Pertaining to the area adjoining an ice-sheet or glacier.

Phenocryst Megascopically visible crystals which stand out prominently from the finer grained ground mass in porphyritic igneous rocks.

Poikilitic An igneous texture in which small crystals are enclosed randomly within larger crystals of another mineral.

Porphyritic An igneous texture describing relatively large grains set in a finer grained matrix.

Pyroclastic Fragmental volcanic rocks created during eruptions.

Rheomorphism A complex process of transformation of rocks into a partially or totally fluid mass.

Rhyolite A fine grained acid volcanic lava, similar in composition to a granite. Frequently contains small phenocrysts.

Sole structures Sedimentary structures formed commonly at the base of sandstones by scouring or sinking into softer muds.

Spherulites Spherical bodies found in acid igneous glasses or acid rocks, generally less than a few centimetres in size. The mineral constituents are arranged in a radial fashion within the bodies.

Stoping The ascent or descent of blocks of the country rock into an enclosing magma.

Subduction zone A zone where ocean floor is destroyed by one lithospheric plate overriding another. The overridden plate is pushed down into the mantle.

Syenite A coarse grained intermediate igneous rock formed mainly of potash feldspars and hornblende. Quartz can be present in small amounts.

Syntectonic plutonism Major igneous activity synchronous with active tectonic movements.

Welded tuff An indurated tuff resembling rhyolite or obsidian in which glassy particles remain sufficiently plastic to weld together (see ignimbrite).

Xenoliths Rock fragments which are foreign to the igneous rock in which they occur.